Advanced Structured Material

Volume 158

Series Editors

Andreas Öchsner, Faculty of Mechanical Engineering, Esslingen University of Applied Sciences, Esslingen, Germany

Lucas F. M. da Silva, Department of Mechanical Engineering, Faculty of Engineering, University of Porto, Porto, Portugal

Holm Altenbach, Faculty of Mechanical Engineering, Otto von Guericke University Magdeburg, Magdeburg, Sachsen-Anhalt, Germany

Common engineering materials reach in many applications their limits and new developments are required to fulfil increasing demands on engineering materials. The performance of materials can be increased by combining different materials to achieve better properties than a single constituent or by shaping the material or constituents in a specific structure. The interaction between material and structure may arise on different length scales, such as micro-, meso- or macroscale, and offers possible applications in quite diverse fields.

This book series addresses the fundamental relationship between materials and their structure on the overall properties (e.g. mechanical, thermal, chemical or magnetic etc.) and applications.

The topics of *Advanced Structured Materials* include but are not limited to

- classical fibre-reinforced composites (e.g. glass, carbon or Aramid reinforced plastics)
- metal matrix composites (MMCs)
- micro porous composites
- micro channel materials
- multilayered materials
- cellular materials (e.g., metallic or polymer foams, sponges, hollow sphere structures)
- porous materials
- truss structures
- nanocomposite materials
- biomaterials
- nanoporous metals
- concrete
- coated materials
- smart materials

Advanced Structured Materials is indexed in Google Scholar and Scopus.

More information about this series at http://www.springer.com/series/8611

Wayne Hall · Zia Javanbakht

Design and Manufacture of Fibre-Reinforced Composites

Springer

Wayne Hall
School of Engineering and Built Environment
Griffith University (Gold Coast Campus)
Southport, QLD, Australia

Zia Javanbakht
School of Engineering and Built Environment
Griffith University (Gold Coast Campus)
Southport, QLD, Australia

ISSN 1869-8433   ISSN 1869-8441   (electronic)
Advanced Structured Materials
ISBN 978-3-030-78809-4   ISBN 978-3-030-78807-0   (eBook)
https://doi.org/10.1007/978-3-030-78807-0

© The Editor(s) (if applicable) and The Author(s), under exclusive license to Springer Nature Switzerland AG 2021
This work is subject to copyright. All rights are solely and exclusively licensed by the Publisher, whether the whole or part of the material is concerned, specifically the rights of translation, reprinting, reuse of illustrations, recitation, broadcasting, reproduction on microfilms or in any other physical way, and transmission or information storage and retrieval, electronic adaptation, computer software, or by similar or dissimilar methodology now known or hereafter developed.
The use of general descriptive names, registered names, trademarks, service marks, etc. in this publication does not imply, even in the absence of a specific statement, that such names are exempt from the relevant protective laws and regulations and therefore free for general use.
The publisher, the authors and the editors are safe to assume that the advice and information in this book are believed to be true and accurate at the date of publication. Neither the publisher nor the authors or the editors give a warranty, expressed or implied, with respect to the material contained herein or for any errors or omissions that may have been made. The publisher remains neutral with regard to jurisdictional claims in published maps and institutional affiliations.

This Springer imprint is published by the registered company Springer Nature Switzerland AG
The registered company address is: Gewerbestrasse 11, 6330 Cham, Switzerland

*To my wife, Angela
and my children, Patrick and Edward*

Wayne Hall

To my parents and my brother

Zia Javanbakht

Preface

There are many excellent composite design and/or manufacturing books in the literature, including those that have a *mechanics of composites* focus. Indeed, many of these books are cited by this text. There are also textbooks that focus on step-by-step practical fabrication methods for making Fibre-Reinforced Composite (FRC) structures—for example, the outstanding 'Composite Materials Fabrication Handbooks 1–3' by Wanberg. So, why write another textbook that focuses on composite materials?

Well... this textbook has resulted from the course '3801ENG Design and Manufacture of Composites' at Griffith University, Australia. This course offers an introduction to FRCs with an emphasis on practical exploration. It aims to span (and link) the content of the aforementioned specialist types of composite textbooks. The purpose of this text is therefore to complement these specialist texts, and to bridge the gap between those with a more design (mechanics) emphasis and those that offer practical fabrication methods.

The course 3801ENG is core in the 'Bachelor of Industrial Design' degree programme, and a free-choice elective for students enrolled in the 'Bachelor of Engineering with Honours in Mechanical Engineering'. The course comprises a 1 hour lecture and 3–5 hours of tutorial activities per week (depending on student requirements), plus individual project-based assessment items that run in parallel with the tutorial activities. Students are assumed to have a basic understanding of *mechanics of materials* concepts (including tension, compression, flexure, transverse shear and torsion), and some previous experience on design and make projects using traditional materials. These skills are included in earlier courses in the Industrial Design programme.

This textbook covers the lecture and tutorial content of 3801ENG. The tutorial classes are used to reinforce the theoretical content covered in the lectures and to establish practical, skill-based competencies in composite fabrication. Practical fabrication methods include wet layup; vacuum bagging; and prepreg moulding. The design and manufacturing examples in this text, including the step-by-step practical fabrication activities, are primarily taken from the tutorial content. Project-based assessment tasks are not included in this textbook—projects are carefully selected, each course offering to link with, as well as build on, the lecture and tutorial content.

Gold Coast, Australia
August 2020

Wayne Hall
Zia Javanbakht

Acknowledgements

The authors would like to acknowledge the help and support provided by our friends and colleagues. A special mention is given to Prof. Andreas Oechsner for his help and encouragement, and to Dr. Ian Underhill for technical support provided in the composite fabrication aspects of this book. The contribution and effort from Dr. Nick Emerson in the intial stages of the textbook is gratefully appreciated. We would also like to acknowledge Prof. John Summerscales for his advice on earlier versions of the manuscript. Finally, we would like to extend our most grateful appreciation to our families. Their continuous support and encouragement have made this book possible.

Contents

1	**Fibre-Reinforced Composites**	1
	1.1 Introduction	1
	1.2 Matrices	2
	1.3 Fibre Types	4
	1.4 Fibre Forms	5
	1.5 Summary	9
	1.6 Questions	10
	1.7 Problems	10
	References	11
2	**Mechanics of Composite Structures**	13
	2.1 Introduction	13
	2.2 Axial (Longitudinal) Modulus	14
	2.3 Transverse Modulus	17
	2.4 Shear Modulus	18
	2.5 Effect of Fibre Orientation	20
	2.6 Strength of Composites	21
	2.7 Woven and Random Fabrics	25
	2.8 Volume Fraction	26
	2.9 Summary	29
	2.10 Questions	30
	2.11 Problems	30
	References	32
3	**How to Make a Composite—Wet Layup**	33
	3.1 Introduction	34
	3.2 Fibre and Matrix Selection	35
	3.3 Mould Preparation	37
	3.4 Layup and Consolidation (Hand Lamination)	39
	3.5 Curing and Post-Curing	42
	3.6 Crosslinking: Chemical Steps	43
	3.7 Demoulding	46
	3.8 Post-Processing (Finishing)	47

	3.9	Summary	49
	3.10	Questions	50
	3.11	Problems	51
	References		51
4	**Advanced Methods—Vacuum Bagging and Prepreg Moulding**		**55**
	4.1	Introduction	55
	4.2	Vacuum Bagging	56
	4.3	Prepreg Moulding	62
	4.4	Summary	64
	4.5	Questions	65
	4.6	Problems	66
	References		67
5	**Composite Testing—How Accurate Are Design Estimates?**		**69**
	5.1	Introduction	69
	5.2	The Tensile Test	70
	5.3	Fibre-Reinforced Composite Specimens	72
	5.4	Tensile Properties: Elastic Moduli and Strength	73
	5.5	Factor of Safety (FoS)	77
	5.6	Summary	77
	5.7	Questions	78
	5.8	Problems	78
	References		79
6	**Moulding Composite Parts**		**81**
	6.1	Introduction	81
	6.2	Composite Part Design	82
	6.3	Mould Design and Construction	85
	6.4	Practical Task: Making a Frisbee	89
	6.5	Summary	100
	6.6	Questions	100
	6.7	Problems	101
	References		103
7	**Hollow Sections—How to Make Composite Tubes**		**105**
	7.1	Introduction	105
	7.2	Mandrel Lamination	106
	7.3	Bladder Moulding	112
	7.4	Practical Task: a Helical Spring	115
	7.5	Extension Task: a Bicycle Handlebar	122
	7.6	Summary	131
	7.7	Questions	131
	7.8	Problems	132
	References		133
Index			**135**

Symbols and Acronyms

Latin Symbols (Capital Letters)

A	Area
E	Elastic modulus
F	Force
G	Shear modulus
I	Second moment of area
J	Polar second moment of area
L	Length
M	Moment (bending)
Q	Shear force
T	Torsional moment (torque)
V	Volume fraction
W	Weight fraction

Latin Symbols (Small Letters)

b	Breadth
d	Diameter
h	Height
n	Number of plies/layers
r	Radius
t	Thickness
y	Distance from the neutral axis

Greek Symbols (Capital Letters)

Δ Denotes the change in a variable.

Greek Symbols (Small Letters)

α Coefficient of thermal expansion
δ Deflection
ϵ Strain
η Fibre efficiency factor
ν Poisson's ratio
ρ Density
σ Tensile or Compressive Stress
τ Shear Stress
θ Angle

Indices (Superscripts)

...*   Strength (or failure)

Indices (Subscripts)

...avg   Average
...b   Bending
...c   Composite
...cl   Composite (longitudinal)
...comp   Compression
...ct   Composite (transverse)
...eq   Equivalent
...f   Fibre
...i   Inner
...m   Matrix
...man   Mandrel
...max   Maximum
...O   Orientation
...o   Outer
...s   Shear

Abbreviations

e.g. for example (from Latin 'exempli gratia')
et al. and others (from Latin 'et alii')
etc. and others (from Latin 'et cetera')
i.e. that is (from Latin 'id est')
viz. namely, precisely (from Latin 'videlicet')
vs. against (from Latin 'versus')

Chapter 1
Fibre-Reinforced Composites

Abstract The concept of fibre-reinforced polymer composites is introduced in this chapter. Most fibre-reinforced composites (FRCs) comprise two constituent materials (known as phases): a fibre phase and a matrix phase. It is noted that thermosetting matrices are far more frequently used for FRCs than their thermoplastic counterparts. An overview of the three most commonly used thermosetting matrices (i.e. unsaturated polyester, vinylester and epoxy) is provided. Available fibre types and forms are considered. The representative mechanical properties for the most common fibre types (glass, carbon and aramid) are presented and compared to traditional materials.

1.1 Introduction

A composite is a material comprising two or more constituent materials. The composite offers a combination of properties that are different from each of the constituents [1, 2]. Many composites comprise only two materials (or phases) [3]: a matrix that surrounds a dispersed phase. In a fibre-reinforced composite (FRC), fibres act as the dispersed phase. Fibres may be long or short [4].

In recent years, there has been a shift towards the use of FRCs. Glass, carbon and aramid fibres are frequently used as reinforcements since they provide high stiffness and strength when used in structural applications. The matrix transfers the load to the fibres, defines the shape and protects the fibres from the environment [1]. Polyester, vinylester and epoxy are commonly used as matrices [5].

Fibre-reinforced composites are now widely used in the transport and energy sectors [6] as well as for numerous other industrial and commercial applications, including storage tanks and pressure vessels [7]. In addition, composites are finding favour in the sports and leisure industry [8] and in the medical sector. Thus, FRCs are now an essential element in industry and technology-based innovation, and therefore an important part of the education of the next generation of mechanical engineering and design graduates [9].

This chapter introduces the fundamentals of fibre-reinforced polymer composites, specifically focusing on the most commonly available fibre reinforcements and matrices.

1.2 Matrices

The polymer matrix of a FRC can be classified as thermoplastic or thermoset [3]. *Thermoplastics* soften (and melt) when heated moderately but return to a solid form upon cooling. *Thermosets* are by far the most commonly used for FRC structures [10]; they are supplied as liquids and tend to offer an easier processing route. They are formed from a chemical crosslinking reaction that is often initiated at ambient temperature. The mixing of the liquid resin and a catalyst (or hardener) causes a non-reversible chemical reaction and a solid polymer is produced. In contrast to thermoplastics, thermosets will not melt upon heating but their mechanical properties may diminish at elevated temperatures. The most common thermosets used for composite applications are polyester, vinylester and epoxy [5].

Polyester. Unsaturated polyester[1] is the most widely used thermosetting resin system (followed by epoxy) [8]. The term unsaturated arises from the carbon double bonds (C=C) in the backbone of the molecular chain [3]. These carbon atoms are not attached to the maximum number of atoms and hence other atoms are able to react at the unsaturated bonds (reactive sites). The unsaturated bonds are therefore the focus of the crosslinking reaction in polyester (as well as in vinylester) [10]. Most polyester resins have a styrene reactive diluent of up to 50% styrene [11] to reduce the resin viscosity (for ease of handling). The styrene molecules also play a significant role in the crosslinking reaction, i.e. they form the links between the polyester chains as polymerisation progresses [10]. Polyesters resins cure in the presence of a catalyst. The catalyst is mixed in small quantities (1–2% is typical [12]) and is usually peroxide-based—for example, methyl ethyl ketone peroxide or MEKP. Polyesters are relatively low-cost and easy to use but they tend to be brittle and prone to high cure shrinkage; a shrinkage of about 7% is typical, but this value could be as high as 12% [13].

Vinylester. Like polyester, vinylester is diluted with styrene and cured with a catalyst [11]. The reactive carbon double bonds (reactive sites) are located at the ends of the molecular chains [8] providing a tougher and more resilient matrix [11] with typically lower shrinkage than polyester [14]—polyester typically shrinks 5–12%, whilst the shrinkage of vinylester is only 5–10% [13]. In addition, the backbone is similar to that of epoxy and this combined with fewer ester groups improves the water and chemical resistance in comparison to its polyester counterpart—ester molecules are susceptible to hydrolysis [11]. Vinylester resins are more expensive than polyester resins and are often harder to source. Moreover, vinylesters tend to be out-performed by their more expensive (and more widely available) epoxy counterparts, particularly in terms of water and chemical resistance and, as a result, they are sometimes overlooked [5].

Epoxy. Epoxy resins are more expensive than either polyester or vinylester resins [3]. They offer higher performance in terms of mechanical and thermal properties, and as

---

[1] There are two main polyester types: orthophthalic and isophthalic resins. Orthophthalic resins are the standard economic (or basic) resins, whilst isophthalic are general purpose resins with a higher cost but superior performance, especially in marine environments [11].

1.2 Matrices

Table 1.1 Advantages, disadvantages and typical mechanical properties of polyester, vinylester and epoxy [3, 19, 20]

Material	ρ ($\frac{g}{cm^3}$)	E (GPa)	σ^* (MPa)	ν	α ($\times 10^{-6} \frac{1}{°C}$)	Advantages	Disadvantages
Polyester	1.04–1.46	2–4.5	34.5–103.5	0.37–0.39	55–200	Lowest cost resin. Low viscosity. Room-temperature curing.	Moderate stiffness and strength. High styrene content. High cure shrinkage.
Vinylester	1.12–1.32	3.0–3.5	73–81		53	Better mechanical properties than polyester. Improved water and chemical resistance. Lower cost than epoxy.	More expensive than polyester. High styrene content. Inferior to epoxy.
Epoxy	1.1–1.4	2.41–4.10	27.6–130	0.38–0.40	45–117	Excellent strength/stiffness. Good thermal stability. Good solvent resistance. Lowest shrinkage.	Highest cost. Poor UV resistance. Post-cure requirements.

previously mentioned, improved water and chemical resistance [10]. The improved water and chemical resistance stems from the lack of ester groups that are evident in polyester and vinylester [11]. They also exhibit lower shrinkage than polyester or vinylester (1.2–4% [4]). The main weakness of epoxies, however, is that they discolour with exposure to UV light. Evidence of UV discolouration is a yellow tinge [15] for clear epoxies. Epoxies are mixed with a hardener to initiate the crosslinking reaction. Typically, an amine hardener is used (co-reaction) to open the epoxy rings (two carbon atoms bonded to an oxygen atom) at either end of the molecular chain [10].

Polyester, Vinylester or Epoxy? There are advantages and limitations of each thermosetting resin type, so the choice of matrix depends largely on the suitability of a resin for a given application. A comparison of various material properties (mass density ρ, modulus of elasticity E, tensile strength σ^*, Poisson's ratio ν, and the coefficient of thermal expansion α along with some of the main advantages and disadvantages of unsaturated polyester, vinylester and epoxy resin are listed in Table 1.1.

1.3 Fibre Types

The most common fibres used as structural reinforcements are glass, carbon and aramid [3, 16]. Other specialist fibres are also commercially available, which include natural fibres such as flax, hemp, jute and kenaf [17, 18], but these are not considered here.

Glass fibres. Glass fibres are usually E-glass (electrical), but other types of glass fibres (for example, S- or structural glass) are also available [19]. The low-cost and balanced mechanical properties of E-glass fibre make it the most common fibre choice [4]. S-glass is stiffer (by approx. 20%) and stronger (30–40% stronger) than E-glass (see Table 1.2), which provides some justification for why it is significantly more expensive.

Carbon fibres. Carbon fibres are identifiably black in colour and are more expensive than either glass or aramid fibres, but tend to offer superior elastic modulus and (at least) comparable strength. Carbon is available in several grades such as standard modulus and high strength carbon (HS), or high modulus carbon (HM). High strength (standard modulus) carbon fibres are the most readily available and tend to be preferred, except for unique or specialist applications.

Table 1.2 Typical mechanical properties for selected fibre types [3, 4, 10, 19, 21]

Material	ρ (g/cm³)	E (GPa)	σ^* (MPa)	ν	α ($\times 10^{-6} \frac{1}{°C}$)	Elongation (%)
Glass (E-glass)	2.50–2.62	69–81	2000–3450	0.22	4.9–5.2	2.6–4.9
Glass (S-Glass)	2.46–2.50	83–89	4585–4800		5.3–5.7	5.7
Carbon (HS*)	1.75–1.80	228–300	3400–7100	0.2	−0.6 (axial), 10 (radial)	1.1–2.4
Carbon (HM†)	1.80–1.95	350–550	1900–4500	0.2	−0.7 (axial), 10 (radial)	0.4–0.7
Aramid (HT‡)	1.40–1.44	83–85	3000–3606			4.0
Aramid (HM)	1.40–1.44	130–131	3600–4100		−6 (axial), 60 (radial)	2.8
Aluminium 6061	2.70	69	124–310	0.33	23.6	17–30
Steel A36 (mild)	7.85	207	400–500	0.30	11.7	23
Titanium Ti-6Al-4V	4.43	110–114	900–1172	0.34	8.6	10–14

*High strength
†High modulus
‡High toughness

1.3 Fibre Types

Aramid fibres. Aramid fibres (from the term *aromatic polyamide*) are yellow or golden fibres with high tensile strengths comparable to carbon fibres [19, 21], but relatively low compressive strength [22]. The term *Kevlar* is sometimes incorrectly used in place of aramid; Kevlar® is the DuPont trade-name. Like glass and carbon, there are also different types of aramids [10]. Aramid fibres have low densities (even lower than carbon) and due to their toughness, they tend to perform very well in impact applications. As a result, aramid fibres are often used in laminates for ballistic applications [4].

Glass, Carbon or Aramid? Similar to the matrix selection, the choice of a fibre type depends on the composite application and requisite material properties. A comparison of the physical, mechanical and thermal properties for a selection of fibre types is provided in Table 1.2. To provide context, some common metals are shown for comparison.

1.4 Fibre Forms

Fibre Bundles. Reinforcement fibres are sometimes supplied as fibre bundles and come wound on a reel. The terms *strand* (or sometimes *end*), *tow* or *roving* may also be used [4]. Strand is used as a generic term and may be applied to glass, carbon or aramid fibres [5]. Tow is often used for high performance carbon and aramid fibres, whilst roving is normally reserved to describe thicker, low-cost glass fibre bundles. The use of strands is normally associated with manufacturing methods such as spray up [10] and filament winding [8, 10], which are beyond the scope of this textbook.

Reinforcement Fabrics. The term *fabric* is used here to describe a fibre assembly that produces a flat sheet in either woven or nonwoven format. Fabrics are usually sold on a roll by the linear metre and are described in terms of fabric weight (areal weight) in grams per square metre (gsm or g/m^2). Narrow fabric strips are often referred to as *tapes*.

Fabrics are characterised here into four main types: unidirectional, woven, braided and random mat. Other fabrics that fall outside these classifications are beyond the scope of this textbook. Fabrics tend to be based on long fibres, but random mats may have a long, continuous swirl of fibres or chopped short fibres (for example, chopped strand mat) [4].

Unidirectional Fibres. A unidirectional (UD) fabric is a nonwoven fabric with fibres mainly orientated in one direction [8]—see Fig. 1.1. Only a small amount of fibres (or a binder material) are orientated perpendicular to the main fibre reinforcements to maintain the integrity of the fabric structure. A unidirectional fabric is easy to handle and can be laid flat and straight in a mould as there are no gaps between fibres. This tends to result in higher fibre volume fractions (i.e. a higher volume of fibres in the composite) and hence, unidirectional fibres are used to produce composites with the maximum stiffness and strength, but this performance is only in the direction of the fibres [11]. Fabrics can be overlapped at multiple orientations to resist the requisite

Fig. 1.1 Unidirectional (UD) fabric

applied loads [19], making UD fabrics an extremely versatile option when designing FRC parts.

Woven Fabrics. Woven fabrics are created by interlacing fibre bundles at right angles to each other. The two orientations are referred to as the warp (0°) and weft (90°) as shown in Fig. 1.2.

Three common woven fabrics are [4]:

- Plain weave.
- Twill weave.
- Satin weave.

In plain weave, each warp strand passes over and then under each weft strand—see Fig. 1.3a. The result is a symmetrical and stable fabric, but a fabric that tends to experience poor *drape* and is more prone to surface porosity (*wet out* issues) than its twill or satin weave counterparts [23]. Thus, plain weave fabrics tend to be preferred for flat, planar laminates or parts with only slight curves [5].

Twill weave fabrics have a balance of stability and drapeabilty so tend to find a wide range of applications. They have lower *crimp* (each fibre bundle follows a less wavy path) than plain weave fabrics and therefore a better drape [10]. They also tend to not experience the same porosity issues sometimes associated with plain weaves. A common twill weave arrangement is the 2/2 pattern; the warp fibre passes over two and then under two weft strands as shown in Fig. 1.3b. This 2/2 twill fabric pattern produces the commonly sought-after carbon fibre aesthetic [11]—see Fig. 1.3c.

Satin or harness weave [10] is the least stable but has good drape and the lowest crimp. Thus, it tends to be mostly used for moulded parts with complex curvature [5] as the lack of stability makes it more difficult to maintain accurate fibre orientations than with the other two woven fabrics. A satin weave is asymmetrical with fibres running predominately in the warp direction on one fabric face as shown in Fig. 1.3d and in the weft on the other face [11]. This asymmetry should be considered when designing composite parts to ensure *coupling stresses* [19] do not cause warping under load.

1.4 Fibre Forms

Fig. 1.2 Warp and weft of a plain weave fabric

Braided Fabrics. Braiding can be used to produce flat tapes [8] and this style is therefore considered to be a fabric [10]. It has some similarity with woven fabrics. Most commonly, however, braiding is used to produce braided sleeves, a specialist fibre preform. Braided sleeves (sometimes referred to as either *socks* or *tubes* [5]) are supplied as a continuous tube of fibres—see Fig. 1.4. Multiple strands are interwoven to create a tubular fabric [8]. Typically, braids are supplied with fibres at ±45° to the tube's longitudinal axis with orientation based on a set diameter. Expanding or compressing the tube's length reduces or increases the diameter with a corresponding change in fibre orientation. Fibre orientations can change between 25 and 75° [11] to accommodate tubular structures with variable diameter [5].

Random Fibres. Random fibres are supplied as either mat or veil—see Fig. 1.5. Random mats may use long fibres (continuous strand mat) or short chopped fibres [4]. They provide a near *uniform* distribution of fibre orientations and therefore produce near isotropic mechanical properties (i.e. elastic modulus and strength) [24]. Glass fibres are extensively used in random mats. The random arrangement of the glass fibres results in large gaps between strands and hence excess resin in the finished laminate [5]. Nonetheless, random mats are cheap and relatively easy to handle, even though (as is the case for chopped strand mat) they may not be as rugged as woven fabrics [10]. They are used to quickly build up the FRC thickness but are rarely used in high performance applications due to the low fibre volume fractions that are usually achieved [11]. They therefore tend to be the chosen reinforcement for composite mould tools (*composite tooling*), lower performance structures [25] and for composite repair.

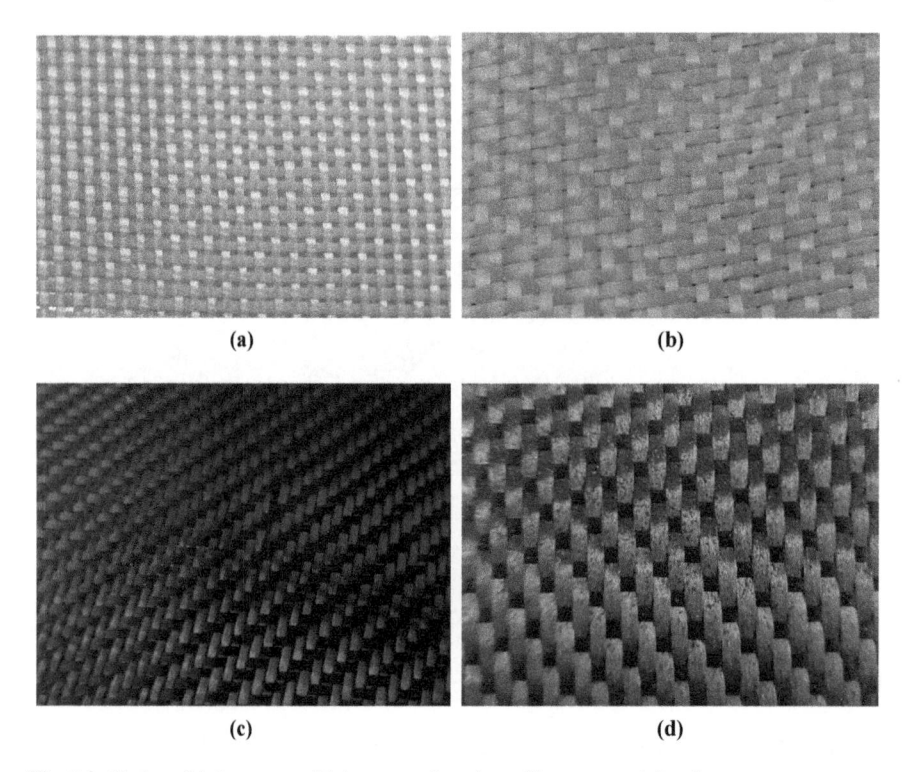

Fig. 1.3 Various fabric types: **a** Plain weave; **b** and **c** twill weave; and **d** satin weave

Fig. 1.4 Braided ±45° carbon sleeve

1.4 Fibre Forms

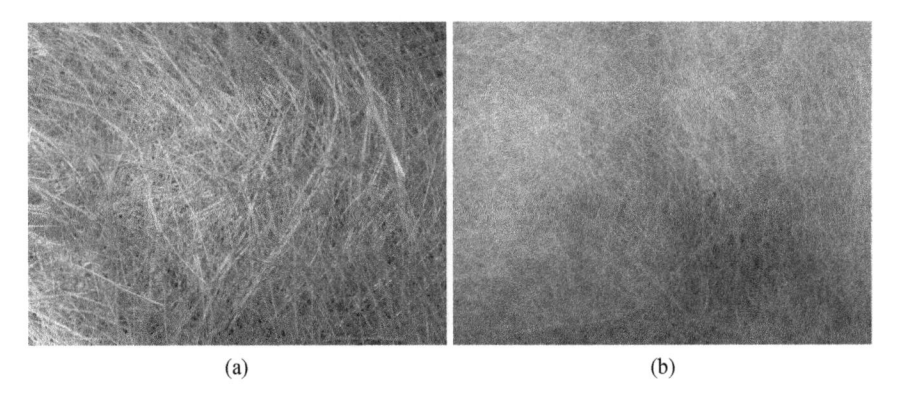

Fig. 1.5 Random fabrics: **a** chopped strand mat; and **b** veil

A tissue or veil is similar to a random mat except that the fibres are much finer. Their primary role is not as a reinforcement, but rather as a surface layer [5]. A veil is often used as a supportive backing for a *gelcoat* [10]; a gelcoat is a resin with pigment and fillers added to provide a high-quality finish that is wear and weather resistant [26].

1.5 Summary

The most common thermosetting resins are:

- Polyester.
- Vinylester.
- Epoxy.

Typical fibre types used as structural reinforcement are:

- Glass.
- Carbon.
- Aramid.

Most fibres are sold as a flat sheet fibre assembly, referred to as a fabric. Fabrics are usually supplied as a roll but narrow fabric strips referred to as tapes can also be purchased. Fabric types can be categories as:

- Unidirectional.
- Woven (common types are plain, twill and satin).
- Braided (including sleeves).
- Random fibres (random mat and veil).

Material properties for some common fibres and matrices are shown in Table 1.3.

Table 1.3 Some common fibre properties

Material	Density (g/cm^3)	Modulus of elasticity (GPa)	Tensile strength (MPa)
Glass (E-Glass)	2.50–2.62	69–81	2000–3450
Carbon (HS)	1.75–1.80	228–300	3000–7000
Carbon (HM)	1.80–1.95	350–550	1900–4500
Polyester	1.04–1.5	2.0–4.5	34.5–103.5
Vinylester	1.12–1.32	3.0–3.5	73–81
Epoxy	1.1–1.4	2.41–4.1	27.6–130

1.6 Questions

Question 1.1 Define the term *composite*. How does the term *composite* differ from *fibre-reinforced composite*?

Question 1.2 What are the main advantages of FRCs?

Question 1.3 In a FRC, define the roles of the matrix and reinforcement fibres.

Question 1.4 Explain the difference between *thermoplastic* and *thermosetting* polymers?

Question 1.5 Identify the three most common thermoset matrices used in composite applications. List the advantages and limitations of each matrix.

Question 1.6 What are the three most common fibre types used to make composites?

Question 1.7 Define the following terms:
a. Strand.
b. Tow.
c. Roving.

Question 1.8 Describe (with the aid of sketches, if necessary) the following fabrics:
a. Unidirectional.
b. Woven.
c. Braided.
d. Random fibres.

1.7 Problems

Problem 1.1 Typical stress-strain curves for E-glass, carbon (HS) and aramid (HT) fibre are shown in Fig. 1.6.

1.7 Problems

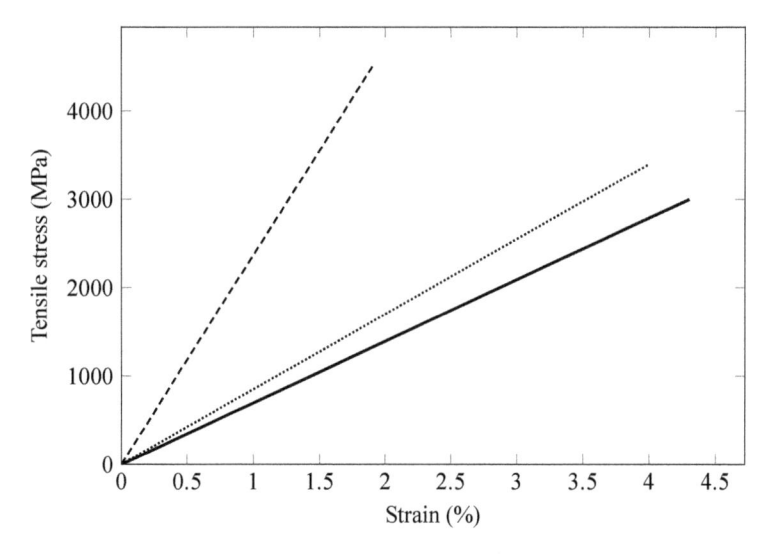

Fig. 1.6 Tensile stress-strain behaviour of a HS carbon-epoxy composite

a. Identify the stress-strain curve for each fibre type.
b. Calculate the elastic moduli of the three fibres from their stress-strain curves.

Answer. a. Carbon, aramid and E-glass (left to right); b. 237 GPa, 85 GPa and 69.7 GPa.

Problem 1.2 From the stress-strain data in Problem 1.1, identify

a. The stiffest fibre.
b. The strongest fibre.
c. The most ductile fibre.
d. The toughest fibre.

Answer. a. Carbon; b. Carbon; c. E-glass; d. Aramid.

References

1. Askeland DR, Fulay PP (2009) Essentials of materials science and engineering, 2nd edn. Cengage Learning, Australia
2. Shackelford JF (2015) Introduction to materials science for engineers, 8th edn. Pearson, Boston
3. Callister WD, Rethwisch DG (2018) Materials science and engineering: an introduction, 10th edn. Wiley, Hoboken NJ
4. Barbero EJ (2017) Introduction to composite materials design, 3rd edn. Composite materials. CRC Press, Boca Raton
5. Wanberg J (2009) Composite materials: fabrication handbook #1, vol 1. Composite garage series. Wolfgang Publications, Stillwater, Minnesota

6. Jones RM (1999) Mechanics of composite materials, 2nd edn. Taylor & Francis, Philadelphia, Pa. and London
7. Vasiliev VV, Morozov EV (2013) Advanced mechanics of composite materials and structural elements, 3rd edn. Elsevier, Amsterdam
8. Astrom BT (2018) Manufacturing of polymer composites, 2nd edn. Routledge, Boca Raton
9. Hall W, Palmer S (2015) Student opportunities in materials design and manufacture: introducing a new manufacturing with composites course. J Mater Educ 37(3–4):155–168
10. Strong AB (2008) Fundamentals of composites manufacturing: materials, methods and applications, 2nd edn. Society of Manufacturing Engineers, Dearborn, Mich
11. Gurit (2019) Guide to composites. https://www.gurit.com/Our-Business/Composite-Materials
12. Fibreglass & Resin Sales (2011) Directions for using MEKP catalyst. https://www.fibreglass-resin-sales.com.au/wp-content/uploads/MEKP-sheet1.pdf
13. Fiore V, Valenza A (2013) Epoxy resins as a matrix material in advanced fiber-reinforced polymer (frp) composites. In: Advanced fibre-reinforced polymer (FRP) composites for structural applications. Elsevier, pp 88–121. https://doi.org/10.1533/9780857098641.1.88
14. Lubin G (ed) (2013) Handbook of Composites. Springer, New York, NY
15. Hollaway L (1990) Polymers and polymer composites in construction. Civil engineering design. Thomas Telford Publishing
16. Smith WF, Hashemi J, Presuel-Moreno F (2019) Foundations of materials science and engineering, 6th edn. McGraw-Hill Education, New York, NY
17. Summerscales J, Dissanayake NP, Virk AS, Hall W (2010) A review of bast fibres and their composites. part 1 – fibres as reinforcements. Compos Part A: Appl Sci Manuf 41(10):1329–1335. https://doi.org/10.1016/j.compositesa.2010.06.001
18. Summerscales J, Virk AS, Hall W (2013) A review of bast fibres and their composites: part 3 - modelling. Compos Part A: Appl Sci Manuf 44:132–139. https://doi.org/10.1016/j.compositesa.2012.08.018
19. Hull D, Clyne TW (1996) An introduction to composite materials, 2nd edn. Cambridge solid state science series. Cambridge University Press, Cambridge
20. Agarwal BD, Broutman LJ, Chandrashekhara K (2006) Analysis and performance of fiber composites, 3rd edn. Wiley and Chichester, Hoboken, N.J
21. Akovali G (2001) Handbook of composite fabrication. Woodhead Publishing, Shawbury
22. Virk AS, Hall W, Summerscales J (2009a) Multiple data set (mds) weak-link scaling analysis of jute fibres. Compos Part A: Appl Sci Manuf 40(11):1764–1771. https://doi.org/10.1016/j.compositesa.2009.08.022
23. Campbell FC (2010) Structural composite materials. ASM International, Materials Park, Ohio
24. Lokensgard E (2010) Industrial plastics: theory and application, 5th edn. Delmar Cengage Learning, Clifton Park NY
25. Fibreglast (2019) The fundamentals of fiberglass. https://www.fibreglast.com/product/the-fundamentals-of-fiberglass/Learning_Center
26. Vaitses AH (2008) The fiberglass boat repair manual. International Marine, Camden, Me

Chapter 2
Mechanics of Composite Structures

Abstract This chapter addresses the mechanical properties of a fibre-reinforced composite (FRC). The focus is on calculation of the elastic modulus and strength for unidirectional FRCs using the rule of mixtures expressions, but woven and random fibres are also considered. A unidirectional FRC exhibits anisotropic behaviour. It is stiffest and strongest in the fibre direction but is relatively compliant and weak in the transverse orientation. Woven structures provide similar mechanical properties in their axial (longitudinal) and transverse orientations, whilst random fibre composites simulate in-plane isotropic behaviour. The effect of the fibre and matrix properties on the structural behaviour of a composite is investigated in the context of fibre volume fractions (and weight fractions), assuming no voids and perfect fibre-matrix adhesion.

2.1 Introduction

The design of a FRC requires:

- The selection of materials (fibres and matrix).
- An appropriate choice of a manufacturing method.

In this chapter, fibre and matrix selection is considered in the context of structural performance. The selection of an appropriate manufacturing method is covered in Chaps. 3 and 4.

Material selection is a complex task as the mechanical properties of FRCs are usually anisotropic, i.e. their mechanical properties are direction-dependent [1]. A fibre-reinforced laminate is made up of several laminae (or plies) stacked on top of each other. Often unidirectional or woven plies are used, but simple low-cost random fibre mats also offer some design and manufacturing opportunities for composite materials. Random mats tend to exhibit near-isotropic (referred to as quasi-isotropic) behaviour [2], and are often used in the marine and automotive sectors for body repair [3].

Herein, an introduction to the mechanics of FRCs is provided to assist material selection. The calculations include many simplifications and assumptions, and should not be considered more accurate than an estimate. Design calculations are an important step in any product design process but, on their own, are not a substitute for physical testing.

2.2 Axial (Longitudinal) Modulus

A fibre-reinforced composite subjected to an axial (tensile or compressive) load will increase or decrease in length. A tensile force will cause an increase in length of the composite, whilst a compressive one will cause a decrease in its length. This change in length depends on several factors including:

- The magnitude of the load.
- Geometrical dimensions of the composite.
- Mechanical properties of the FRC material.

The mechanical properties of a fibre-reinforced composite are influenced by the properties of both the fibres and matrix; the amount (usually referred to as the volume fraction) of the fibres; the fibre length and the orientation of the fibres; and the fibre-matrix adhesion characteristics [2, 4, 5].

Now, consider a fibre composite made up of a number of parallel unidirectional plies stacked on top of each other to create a solid, homogenous composite material. This FRC is shown in Fig. 2.1 under tension and compression. The plies consist of long, continuous fibres in a matrix. The fibres are all orientated in the same direction, referred to as the axial [5] or longitudinal [2, 4] direction. When all fibres are orientated in the same direction, the composite is said to be an aligned fibre composite, a unidirectional composite or simply a UD composite.

The mechanical properties of the composite material are anisotropic; a unidirectional composite has the highest elastic modulus (stiffness[1]) and strength in the fibre (longitudinal) direction. The stiffness and strength of a unidirectional composite perpendicular (often termed transverse) to the fibre is much lower [5]. The fibre and matrix contributions govern the mechanical properties of the longitudinal and transverse directions, respectively.

A typical tensile stress-strain curve for a high-strength (HS) carbon-epoxy unidirectional composite loaded parallel to the fibres is shown in Fig. 2.2. The stress-strain curves for the stiff and strong carbon fibres, and the weaker but more ductile epoxy matrix are overlaid for comparison. The composite performance lies somewhere between that of the fibres and matrix behaviour. Tensile stress is proportional to strain, and the elastic modulus (or Young's modulus of elasticity) is the constant of proportionality between the stress and strain [1]. The tensile strength of the FRC is

---

[1] Note, the terms *stiffness* and *elastic modulus* are sometimes used interchangeably in this text. In fact, stiffness is not the same as elastic modulus but they *are* related. This stiffness and elastic modulus relationship is dependent on specimen dimensions and the load application. For instance, in axial members, E is actually related to stiffness (measured along the member's length) via EA/L.

2.2 Axial (Longitudinal) Modulus

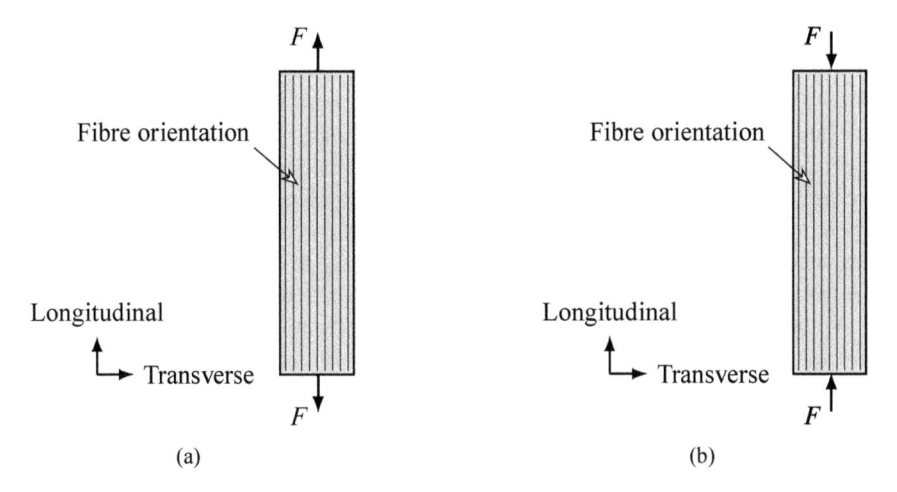

Fig. 2.1 Fibre composite members subjected to axial load: **a** tension; and **b** compression

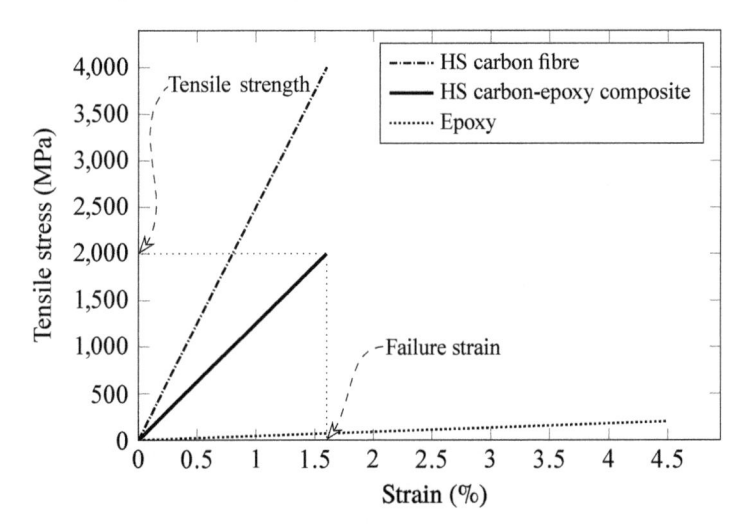

Fig. 2.2 Tensile stress-strain behaviour of a HS carbon-epoxy composite

the maximum stress measurement, whilst the failure strain is the strain at composite fracture, corresponding here to the strain at the maximum stress value.

The longitudinal modulus of the composite, E_{cl}, is calculated by the rule of mixtures (RoM) [1], viz.

$$E_{cl} = E_m(1 - V_f) + E_f V_f \qquad (2.1)$$

where E_f and E_m are the elastic moduli of the fibres and matrix, respectively, and V_f is the fibre volume fraction.

Note that tensile and compressive stiffness measurements of the composite are found to be similar, but some minor discrepancies exist, e.g. a difference in tensile modulus of only about 10% is observed for carbon-epoxy composites [6]. Thus, Eq. (2.1) is relevant under both tension and compression load cases.

The elastic moduli data for common fibres and matrices are readily available in the literature and representative values have been presented in Tables 1.1 and 1.2. These material properties are provided to enable the calculation of design estimates (i.e. mechanical properties) for FRCs, and ultimately to size the composite. Note, the term *size* is used here to mean 'identify the requisite number of plies'.

Volume fractions are used to describe the proportion of fibres in the composite structure. The upper *theoretical* or threshold fibre volume fractions are considered later (see Sect. 2.8), whilst sensible (practically attainable) fibre volume fraction estimates for each manufacturing process are considered in Chaps. 3 and 4.

Example 2.1

A unidirectional composite consists of glass fibres ($E_f = 69$ GPa) embedded in a polyester matrix ($E_m = 2.8$ GPa). Assume $V_f = 0.4$. Calculate the longitudinal elastic modulus for the composite.

Solution

The longitudinal modulus is computed from

$$E_{cl} = E_m(1 - V_f) + E_f V_f$$
$$= 3 \times 10^9 (1 - 0.4) + 69 \times 10^9 \times 0.4$$
$$= 29.3 \times 10^9 \text{ N/m}^2 = 29.3 \text{ GPa}$$

Note. E_{cl} is *fibre-dominated*; the fibres support nearly all of the applied load with only a very minor matrix contribution [1]. Thus, we can simplify the RoM calculation further by using *only* the fibre characteristics to offer a sensible (albeit, a slightly lower) estimate of E_{cl} for the unidirectional laminate.

Thus

$$E_{cl} \approx E_f V_f$$
$$\approx 69 \times 10^9 \times 0.4$$
$$\approx 27.6 \times 10^9 \text{ N/m}^2 = 27.6 \text{ GPa}$$

The obtained value is similar but slightly lower (<10%) than the previous calculation of 29.3 GPa.

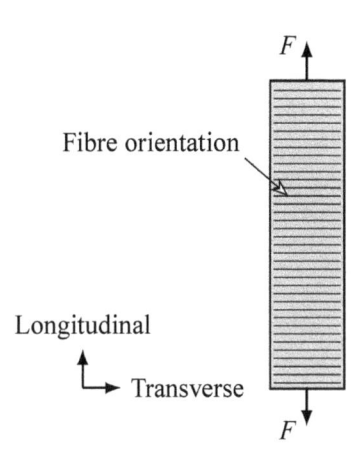

2.3 Transverse Modulus

Fig. 2.3 Composite member with fibres orientated perpendicular to the tensile load

So far, we have only considered composite members with fibres orientated parallel (along the length of the member) to the applied load. Now consider the situation where the fibres are orientated perpendicular to the loading direction in the transverse direction—see Fig. 2.3.

In this situation, it is the matrix (rather than the fibres) that dominates the stiffness response, and the transverse modulus E_{ct} is given by [1][2]

$$\frac{1}{E_{ct}} = \frac{1 - V_f}{E_m} + \frac{V_f}{E_f} \qquad (2.2)$$

The transverse modulus expression in Eq. (2.2) is sometimes referred to as the inverse rule of mixture (IRoM) [8, 9]. Whilst the RoM expression in Eq. (2.1) assumes the internal strains in the fibres and matrix are equal, referred to as iso-strain (or sometimes the *Voigt model* [10]), this IRoM (Eq. (2.2)) assumes the internal stresses are the same in the fibres and matrix, termed iso-stress or sometimes the *Reuss model* [11].

> **Example 2.2**
>
> A unidirectional carbon fibre-reinforced epoxy composite has a fibre volume fraction of 0.5. Calculate the elastic modulus of the composite in the longitudinal and transverse directions. Assume the moduli of elasticity of the carbon fibre reinforcements and the epoxy matrix are 230 GPa and 4 GPa, respectively.

---

[2] A significant underestimate is known to result from this simple expression but it is considered here (in this introductory text) to offer a sensible design estimate. If necessary, a more accurate estimate can be calculated using the more complex, semi-empirical Halpin-Tsai expression in [7].

Solution

The longitudinal modulus is calculated from

$$E_{cl} = E_m(1 - V_f) + E_f V_f$$
$$= 4 \times 10^9 (1 - 0.5) + 230 \times 10^9 \times 0.5$$
$$= 117 \times 10^9 \text{ N/m}^2 = 117 \text{ GPa}$$

The transverse modulus is calculated from

$$\frac{1}{E_{ct}} = \frac{(1 - V_f)}{E_m} + \frac{V_f}{E_f}$$
$$\frac{1}{E_{ct}} = \frac{(1 - 0.5)}{4 \times 10^9} + \frac{0.5}{230 \times 10^9}$$
$$\frac{1}{E_{ct}} = 0.127 \times 10^{-9}$$
$$E_{ct} = 7.9 \times 10^9 \text{ N/m}^2 = 7.9 \text{ GPa}$$

2.4 Shear Modulus

Figure 2.4 shows a FRC subjected to in-plane shear. The shear modulus is the ratio of the shear stress, τ, to the shear strain, γ [1]. Shear stress and shear strain are proportional, and the shear modulus (like the elastic modulus for the tensile and compressive load cases) is the constant of proportionality [12]. The shear modulus of the composite, G_c, can be estimated using a similar expression to the transverse modulus [5], via

$$\frac{1}{G_c} = \frac{1 - V_f}{G_m} + \frac{V_f}{G_f} \quad (2.3)$$

where G_f and G_m are the shear moduli of the reinforcing fibres and matrix, respectively.

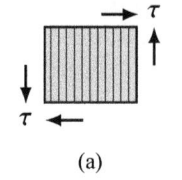

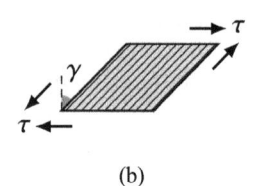

Fig. 2.4 A unidirectional element subjected to in-plane shear stress: **a** undeformed; and **b** deformed

2.4 Shear Modulus

Example 2.3

A unidirectional E-glass fibre-reinforced epoxy composite is required to have an axial modulus of more than 35 GPa, and a shear modulus of at least 2.5 GPa. Table 2.1 provides the mechanical properties for the E-glass fibres and epoxy, respectively. What is the minimum fibre volume fraction that is needed?

Table 2.1 Fibre and matrix properties

Material	Elastic modulus (GPa)	Shear modulus (GPa)
E-glass	70	28.7
Epoxy	3.2	1.2

Solution

First, consider the composite's longitudinal modulus

$$E_{cl} = E_m(1 - V_f) + E_f V_f$$
$$35 \leq 3.0(1 - V_f) + 70 V_f$$
$$\therefore V_f \geq 0.48$$

A volume fraction greater than 0.48 is acceptable to satisfy the axial stiffness requirement.

Now, let's consider the shear modulus

$$\frac{1}{G_c} = \frac{1 - V_f}{G_m} + \frac{V_f}{G_f}$$
$$\frac{1}{2.5} \geq \frac{1 - V_f}{1.2} + \frac{V_f}{28.7}$$

$$\therefore V_f \geq 0.55$$

The minimum volume fraction to satisfy the shear stiffness requirement is 0.55.

Thus, to satisfy both design conditions, a fibre volume fraction, V_f, of at least 0.55 is needed (i.e. a minimum V_f of 0.48 and 0.55 for axial and shear stiffness, respectively).

2.5 Effect of Fibre Orientation

Of course, fibre-reinforced composites are not always loaded either parallel or perpendicular to the fibres, for example, see Fig. 2.5. It is therefore necessary to briefly consider the effect of loading angle on the elastic and shear moduli of the composite.

The effect of fibre orientation angle θ on the elastic and shear moduli of a FRC is shown in Fig. 2.6. The longitudinal (0°) and transverse (90°) elastic moduli were previously calculated in Example 2.2, whilst the in-plane shear modulus has been calculated from Eq. (2.3) assuming typical fibres and matrix values. The *off-axis* elastic and shear moduli values are calculated based on reference [5]; note that the calculation of off-axis elastic and shear moduli values requires an understanding of the Poisson effects. In Fig. 2.6, ν_c is Poisson's ratio for the FRC material defined for a strain applied in the longitudinal direction [5].

The off-axis elastic modulus $E_{c\theta}$ (see Fig. 2.6) initially diminishes rapidly from the longitudinal E_{cl} ($E_{c\theta}$, $\theta = 0°$) value as the fibre orientation angle θ increases. At an angle of 10°, the off-axis elastic modulus has reduced to almost half the longitudinal modulus. The rate of decay of the elastic modulus reduces as the fibre angle increases further but at 45°, the magnitude approaches the same modulus as that for transverse loading ($E_{c\theta}$, $\theta = 90°$); a magnitude of about 10% of the maximum is typical for carbon fibre composites [13]. In contrast, the shear modulus $G_{c\theta}$ is a minimum at an angle of 0 and 90°, and maximum at 45°. So, put simply, in designing optimised composite structures or parts, aligned fibres ($E_{c\theta}$, $\theta = 0°$) are recommended to resist tensile or compressive loads whilst fibres should be orientated at ±45° for shear loads.

The maximum shear modulus can be calculated from stress and strain transformations [5] to be

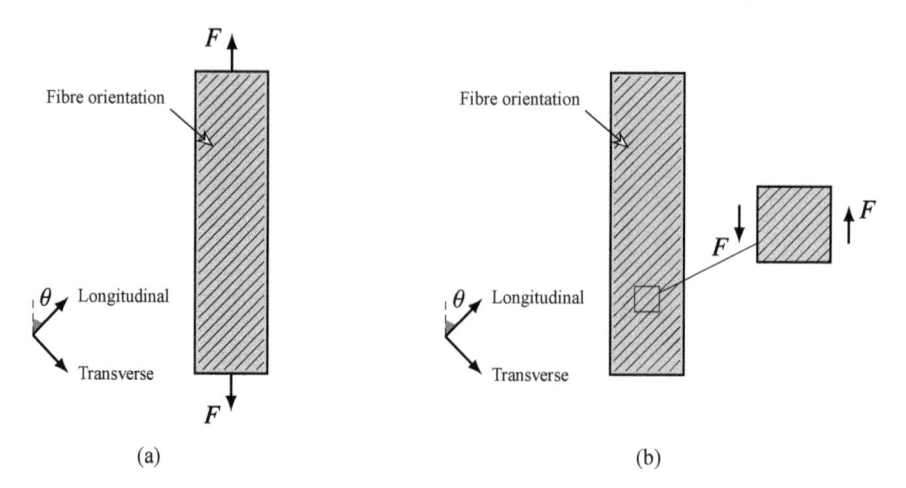

Fig. 2.5 Fibre-reinforced composite member with fibres orientated at an angle θ: **a** tensile loading; and **b** in-plane shear

2.5 Effect of Fibre Orientation

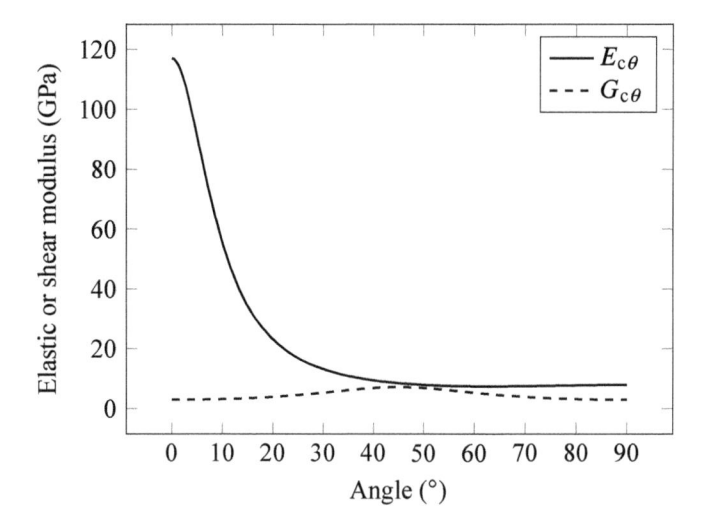

Fig. 2.6 Effect of loading angle on elastic and shear moduli of a unidirectional carbon-epoxy composite ($E_{cl} = 117$ GPa, $E_{ct} = 7.9$ GPa, $G_c = 2.95$ GPa and $\nu_c = 0.265$)

$$\frac{1}{G_c 45°} = \frac{1 + 2\nu_c}{E_{cl}} + \frac{1}{E_{ct}} \tag{2.4}$$

where ν_c can be calculated from a modified version of the RoM equation (see Eq. 2.1).

2.6 Strength of Composites

A composite will fail when the normal or shear stress exceed the strength values. There are three obvious failure modes:

- Axial (longitudinal) failure.
- Transverse failure (of the matrix or fibre-matrix interface).
- Shear failure (of the matrix due to parallel shear forces).

These failure modes are shown schematically in Fig. 2.7. In the figure, the axial tensile and transverse failures are shown for tensile loading only.

Axial (Longitudinal) Failure.
The longitudinal strength of a unidirectional composite depends on whether the applied load is tensile or compressive. If the load is tensile, the Kelly-Tyson (KT) model [14] (which is based on the RoM expression) can be used to predict composite failure. Measured compression longitudinal strengths, however, are usually lower than those noted in tension [15]—often equating to 50–60% [16] (or less [4]) of

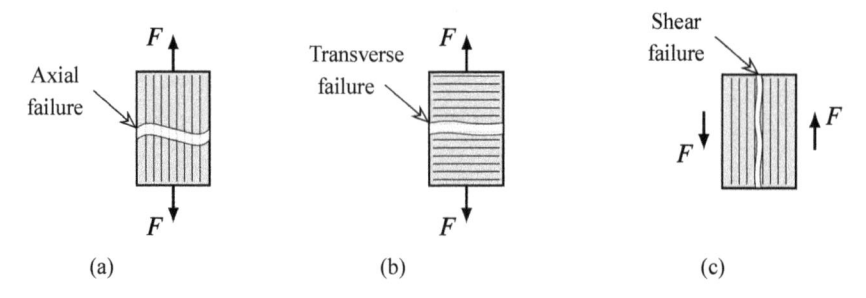

Fig. 2.7 Fibre-reinforced composite failure modes: **a** axial tensile failure; **b** transverse tensile failure and **c** in-plane shear failure

the tensile strength. The failure mechanism is the main reason for this behaviour; microbuckling of fibres govern the longitudinal strength under compression [4].

In axial tension, if we assume the *fibres fail first* (prior to the matrix), the longitudinal strength of the FRC σ_{cl}^* is defined by the Kelly-Tyson (KT) model [14] as

$$\sigma_{cl}^* = \sigma_{m(f^*)}(1 - V_f) + \sigma_f^* V_f \qquad (2.5)$$

where σ_f^* is the failure strength of the fibres and $\sigma_{m(f^*)}$ is the stress in the matrix at fibre failure.

If the *matrix fails first*, it can be assumed that the fibres continue to resist the applied tensile load after the matrix has failed; the matrix therefore provides no contribution and the failure strength is [5]

$$\sigma_{cl}^* = \sigma_f^* V_f \qquad (2.6)$$

The matrix contribution to the tensile strength is small in comparison to the fibre contribution and therefore it follows that the axial (longitudinal) strength of the composite, irrespective of whether the fibres or matrix fail first, will be fibre-dominated.

Example 2.4

The unidirectional carbon fibre-epoxy composite in Example 2.2 is axially loaded in tension until failure. The tensile strength of the HS carbon fibres and the epoxy matrix is 4000 MPa and 65 MPa, respectively. Calculate the failure strength of the composite assuming the stress in the epoxy matrix is 30 MPa when fibre failure occurs.

2.6 Strength of Composites

Solution

The longitudinal tensile strength is calculated from

$$\sigma_{cl}^* = \sigma_{m(f^*)}(1 - V_f) + \sigma_f^* V_f$$
$$= 30 \times 10^6 (1 - 0.5) + 4000 \times 10^6 \times 0.5$$
$$= 2015 \times 10^6 \text{ N/m}^2 = 2015 \text{ MPa}$$

If the matrix stress at fibre failure was unknown, we could conservatively neglect the matrix contribution and still provide a reasonable estimate of the failure strength, via

$$\sigma_{cl}^* = \sigma_f^* V_f$$
$$= 4000 \times 10^6 \times 0.5 = 2000 \times 10^6 \text{ N/m}^2 = 2000 \text{ MPa}$$

This value is close to the 2015 MPa previously calculated when including the matrix stress. Neglecting the matrix contribution provides a result that is akin to a matrix-fail-first scenario. Put simply, the matrix contribution is minimal.

Transverse Failure. Transverse to the fibres, the internal stresses in the fibre-reinforced composite will be the same in the fibres and matrix, and hence transverse failure is likely to occur in the weaker matrix (or often at the fibre-matrix interface). In contrast to the axial strength, similar (relatively low) strength values are noted transverse to the fibres, irrespective of whether the load is tensile or compressive [5]. Most importantly, the strength of the composite will not exceed the matrix strength. Often, however, the measured strength will be significantly less than that of the matrix, viz.

$$\sigma_{ct}^* \ll \sigma_m^* \tag{2.7}$$

Shear Failure. There is no simple analytical expression for estimating the shear strength of a unidirectional composite based on the fibre volume fraction [5]. In the absence of empirical data, the (in-plane) shear strength of a unidirectional composite can (for V_f values below 0.7) conservatively be considered to be approximately equal to that of the matrix, i.e. $\tau_c^* \approx \tau_m^*$ [17]. Moreover, using the Tresca criteria, we can relate the shear strength of the matrix to its tensile strength via [18, 19]

$$\tau_c^* \approx \tau_m^* = 0.5 \times \sigma_m^* \tag{2.8}$$

Failure Under Off-Axis Loading. Evaluating the strength of a UD composite subjected to off-axis loading ($0° < \theta < 90°$) is more complex than for the case where fibres are orientated either parallel or perpendicular to the applied loads; a detailed description is therefore beyond the scope of this text. The complexity arises from an interaction between the normal and shear components [20]. In general, it is reasonable to assume that fibre orientations influence normal and shear strengths in a similar way to the elastic and shear moduli. The maximum tensile or compressive strengths occurs when the fibres are aligned parallel to the normal load (i.e. at 0°), whilst the maximum shear performance occurs with fibre orientations at ±45° to the shear forces. So, whilst fibres are most often aligned in a composite structure at 0° and 90° to resist longitudinal and transverse loads, respectively, some fibres are orientated at ±45°. These ±45° fibres offer minimal resistance to longitudinal and transverse loads, but are much better at resisting shearing and/or torsional loads [13].

Example 2.5

E-glass, S-glass and carbon fibre are available for the design of a unidirectional reinforced epoxy composite. The composite must have a longitudinal modulus of at least 50 GPa and a strength of more than 1000 MPa. Mechanical properties for each of the fibre types and the epoxy are given in Table 2.2. Assuming a maximum fibre volume fraction of 0.5 can be achieved, which of the fibres are suitable?

Table 2.2 Mechanical properties of the constituents (fibres and matrix)

Material	Elastic modulus (GPa)	Tensile strength (MPa)
Carbon fibre	230	4900
E-glass	70	2500
S-glass	95	4600
Epoxy matrix	3.2	60

Solution

Based on the values calculated in Table 2.3, only carbon fibre satisfies both criteria.

Table 2.3 Mechanical properties of the fibre-reinforced composites

Material	Elastic modulus check $E_{cl}=E_m(1-V_f)+E_f V_f$	Tensile strength check $\sigma_{cl}^* = \sigma_{m(f')}(1-V_f) + \sigma_f^* V_f \approx \sigma_f^* V_f$
Carbon fibre	116.6 GPa ✓	2450 MPa ✓
E-glass	36.6 GPa ✗	1250 GPa ✓
S-glass	49.1 GPa ✗	2300 GPa ✓

2.7 Woven and Random Fabrics

So far, the focus of this chapter has been on unidirectional fibre reinforcements but many laminates also incorporate woven fabric into their structure (e.g. modern skis [1]). Some composite structures even incorporate randomly orientated fibres, especially for lower stress applications [21]. So, how can we consider these other fabric forms?

Woven Fabrics. A woven ply contains fibres orientated in both the warp and weft directions (i.e. perpendicular to each other) [3]—see Fig. 1.2. If half of the fibres are orientated in the warp direction and the other half in the weft orientation, V_f is half the total fibre volume fraction in each orientation; this is sometimes referred to as a *balanced* fabric. To provide a simple approximation of the elastic moduli for balanced woven cloths, similar stiffnesses can therefore be assumed in both the warp and weft directions. The elastic modulus values are then calculated using the rule of mixtures approximation from Eq. (2.1) with a fibre orientation efficiency factor, η_o, of ½ included, via

$$E_c = E_m(1 - V_f) + \frac{1}{2} E_f V_f \tag{2.9}$$

Similarly, the tensile strength can be approximated from Eq. 2.5 using the fibre orientation efficiency factor. The shear modulus and in-plane shear strength of a woven composite will be somewhat comparable to a UD laminate [18] since the shear performance of a UD composite at either 0 or 90° orientations is the same—see Fig. 2.6.

Random Mats. Since random fabrics have fibres orientated in all directions, their in-plane elastic modulus can be considered to be the same in all orientations (quasi-isotropic [22]). In other words, random fibre composites can be considered to exhibit isotropic elastic properties. The limitation with this fibre form is that the elastic modulus is usually lower than for the other aligned fibre forms. Moreover, fibre volume fraction tends to be lower due to packing issues. An orientation efficiency of ⅜ [23] is frequently used, yielding

$$E_c = E_m(1 - V_f) + \frac{3}{8} E_f V_f \tag{2.10}$$

Example 2.6

A glass fibre-reinforced laminate comprises 40 vol% woven fibres in an epoxy matrix. The glass fibres and epoxy matrix have elastic moduli of 69 GPa and 3.5 GPa, respectively. The tensile strength of the glass fibres is 3450 MPa. Calculate the elastic modulus and tensile strength in the warp and weft directions.

Solution

Assuming a balanced fabric the warp and weft moduli are calculated from

$$E_c = E_m(1 - V_f) + \frac{1}{2}E_f V_f$$

$$= 3.5 \times 10^9 (1 - 0.4) + \frac{1}{2} \times 69 \times 10^9 \times 0.4$$

$$= 15.9 \times 10^9 \text{ N/m}^2 = 15.9 \text{ GPa}$$

and the warp/weft strengths are estimated from

$$\sigma_c^* = \sigma_{m(f^*)}(1 - V_f) + \frac{1}{2}\sigma_f^* V_f$$

Either the fibres or the matrix may fail first. This will depend upon the failure strain of the constituents. Since the failure mode is not known, we can conservatively neglect the matrix contribution. Remember, even if the fibre fails first, the matrix contribution is likely to be relatively small and will result in a slightly lower failure strength. An underestimate of the composite's strength is better than an overestimate. Thus, conservatively we calculate the warp and weft strengths as

$$\sigma_c^* = \frac{1}{2} \times 3450 \times 10^6 \times 0.4$$

$$\sigma_c^* = 690 \times 10^6 \text{ N/m}^2 = 690 \text{ MPa}$$

2.8 Volume Fraction

The calculation of elastic properties for a unidirectional fibre composite requires knowledge of the fibre and matrix properties, as well as fibre volume fraction data. In contrast, manufacturers usually supply *dry* fibres based on their weight (referred to as areal weight; mass per unit area of a single ply), whilst prepreg suppliers often specify an additional weight fraction of the matrix (or resin weight fraction [24]). It is therefore essential to be able to convert these resin weight fractions into the requisite fibre volume fractions needed for the above design estimates (i.e. for mechanical property calculations).

An understanding of the mass density (mass per unit volume) of the fibres, ρ_f, and matrix, ρ_m, is needed to be able to calculate fibre volume fractions from corresponding weight fractions, W_f and W_m, and vice versa. Typical density values for matrices and

2.8 Volume Fraction

fibres were presented in Tables 1.1 and 1.2, respectively. The fibre volume fraction can be calculated from the weight fraction of the fibres, via [5]

$$V_f = \frac{\frac{W_f}{\rho_f}}{\frac{W_f}{\rho_f} + \frac{1-W_f}{\rho_m}} \qquad (2.11)$$

In a similar manner, the fibre weight fraction can be calculated from the volume fraction as [2, 5]

$$W_f = \frac{V_f \rho_f}{V_f \rho_f + (1-V_f)\rho_m} \qquad (2.12)$$

Example 2.7

A randomly orientated glass fibre-reinforced polyester composite has a fibre weight fraction of 0.4 (i.e. 40 wt%). The density of the glass fibres and the matrix are 2550 kg/m³ and 1130 kg/m³, respectively. Calculate the fibre volume fraction.

Solution

The fibre volume fraction is readily calculated from

$$V_f = \frac{\frac{W_f}{\rho_f}}{\frac{W_f}{\rho_f} + \frac{1-W_f}{\rho_m}} = \frac{\frac{0.4}{2550}}{\frac{0.4}{2550} + \frac{1-0.4}{1130}} = 0.23$$

Example 2.8

A unidirectional HS carbon fibre-reinforced epoxy laminate has a fibre weight fraction of 0.65. Estimate the mechanical properties (i.e. elastic modulus and strength) of the carbon-epoxy composite under longitudinal tensile loading. The fibre and matrix properties are shown in Table 2.4.

Table 2.4 Material properties of the carbon fibres and epoxy matrix

Material	Density, ρ (g/cm³)	Elastic modulus, E (GPa)	Tensile strength, σ^* (MPa)
Carbon fibre	1.76	230	4100
Epoxy matrix	1.18	3.2	55

Solution

The fibre volume fraction is calculated from

$$V_\text{f} = \frac{\frac{W_\text{f}}{\rho_\text{f}}}{\frac{W_\text{f}}{\rho_\text{f}} + \frac{1-W_\text{f}}{\rho_\text{m}}} = \frac{\frac{0.65}{1760}}{\frac{0.65}{1760} + \frac{1-0.65}{1180}} = 0.55$$

The longitudinal elastic modulus is calculated from

$$\begin{aligned} E_\text{cl} &= E_\text{m}(1 - V_\text{f}) + E_\text{f} V_\text{f} \\ &= 3.2 \times 10^9 (1 - 0.55) + 230 \times 10^9 \times 0.55 \\ &= 128.0 \times 10^9 \text{ N/m}^2 = 128.0 \,\text{GPa} \end{aligned}$$

Assuming the fibres fail first, the longitudinal strength is calculated from

$$\sigma_\text{cl}^* = \sigma_{m(\text{f}^*)}(1 - V_\text{f}) + \sigma_\text{f}^* V_\text{f}$$

Since the stress in the matrix at fibre failure is not known, we can conservatively neglect the matrix contribution, hence

$$\sigma_\text{cl}^* = 4100 \times 10^6 \times 0.55 = 2255 \times 10^6 \text{ N/m}^2 = 2255 \,\text{MPa}$$

Packing Arrangement. So far, we have used a relatively narrow range of typical fibre volume fractions. It should be evident that there is a limit to the volume of fibres that can be embedded into a composite. This limit is due to geometrical constraints that result from the size and shape of the fibres, the packing arrangement and the need for a surrounding matrix to transfer the load to the stiff and strong fibres. So, what is this threshold (upper) limit?

Let us start by looking at two packing arrangements: a hexagonal and a simple square arrangement of fibres—see Fig. 2.8. In the figure, the fibre cross-sections are represented by circles and are assumed to have a uniform fibre diameter along their length. All fibres have the same diameter and therefore the distance between the fibre centres (s) in the arrangement must be separated by at least twice the fibre radius.

If we consider the square and hexagonal arrangements, we can show that the square unit cell has a maximum fibre volume fraction of 0.785 whilst that of the hexagonal arrangement is optimised at 0.907. The hexagonal unit cell offers the maximum possible fibre volume fraction (and hence, the upper threshold limit). In practice, however, several factors can influence fibre volume fractions, making these theoretical values unlikely to be attained when manufacturing a FRC. The most important factors include fibre form and the manufacturing method. Table 2.5 offers sensible fibre volume fraction target ranges for unidirectional, woven and random fibre mats.

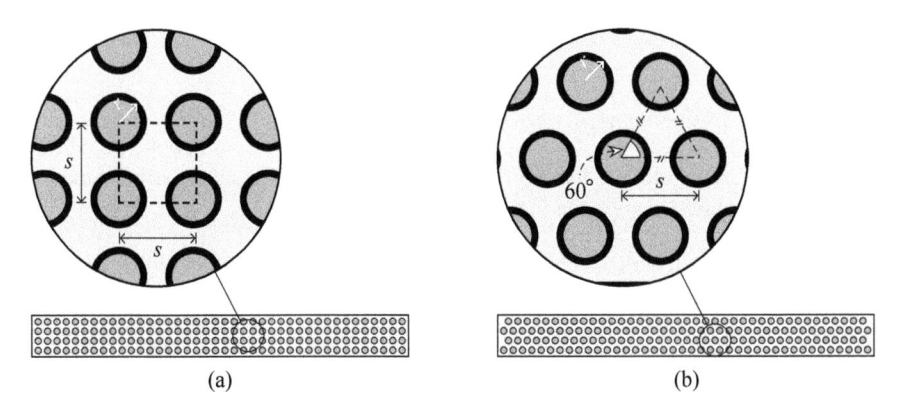

Fig. 2.8 Fibre packing arrays: **a** square arrangement; and **b** hexagonal arrangement

Table 2.5 Practical fibre volume fractions for random, woven and unidirectional composites

Fibre form	Volume fraction (V_f)
Random	0.1–0.3
Woven	0.2–0.55
Unidirectional	0.3–0.7

2.9 Summary

Table 2.6 provides a summary of formulae for estimating the mechanical properties of FRCs.

Table 2.6 Design formulae for fibre-reinforced composites

Composite Type	Property	Formula	
Unidirectional	Longitudinal elastic modulus	E_{cl}	$= E_m(1 - V_f) + E_f V_f$
	Transverse elastic modulus	$\frac{1}{E_{ct}}$	$= \frac{1-V_f}{E_m} + \frac{V_f}{E_f}$
	Longitudinal failure strength	σ_{cl}^*	$= \sigma_{m(f^*)}(1 - V_f) + \sigma_f^* V_f \approx \sigma_f^* V_f$
	Transverse failure strength	σ_c^*	$= \sigma_m^*$
	In-plane shear strength	τ_c^*	$= \tau_m^*$
Woven	Elastic modulus	E_c	$= E_m(1 - V_f) + \frac{1}{2} E_f V_f$
Random mat	Elastic modulus	E_c	$= E_m(1 - V_f) + \frac{3}{8} E_f V_f$

2.10 Questions

Question 2.1 Define the term *anisotropic*.

Question 2.2 How should the fibre reinforcements be orientated to produce the stiffest and strongest composite in tension and compression?

Question 2.3 The tensile strength of a unidirectional composite is higher than the compressive strength. The compressive strength is often what proportion of the tensile strength?

Question 2.4 How should the fibre reinforcements be orientated to produce the stiffest and strongest composite in shear?

Question 2.5 A *fibre orientation efficiency factor* is used to approximate the elastic moduli of woven and randomly orientated fibres. Why?

Question 2.6 Practically, what fibre volume fraction ranges can be achieved for unidirectional, woven and random fibre composites?

Question 2.7 Fibre volume fractions for unidirectional composites are normally higher than their random or woven counterparts. Why?

Question 2.8 Why is it important to be able to convert weight fractions into fibre volume fractions and vice versa?

2.11 Problems

Problem 2.1 An aligned FRC consists of E-glass fibres in an epoxy matrix. The volume fraction of E-glass fibres is 0.5. The fibres have a modulus of 69 GPa whilst the epoxy has a modulus of 3.8 GPa. Calculate the composite's axial (longitudinal) and transverse elastic moduli.
Answer: 36.4 GPa, 7.2 GPa.

Problem 2.2 Determine the elastic moduli (longitudinal and transverse) of a unidirectional carbon fibre-reinforced epoxy composite with 50 vol% fibres. Assume that the fibres and epoxy have an elastic moduli of 230 GPa and 3.8 GPa, respectively.
Answer. 116.9 GPa, 7.5 GPa.

Problem 2.3 A random FRC consists of 30 vol% glass fibres in a polyester matrix. Again, the E-glass fibres can be assumed to have an elastic modulus of 69 GPa whilst the polyester has an elastic modulus of 3.0 GPa. Calculate the in-plane elastic modulus of the composite.
Answer. 9.9 GPa.

2.11 Problems

Table 2.7 Material properties for the glass-polyester composite

Material	Elastic modulus (GPa)	Tensile strength (MPa)
E-glass	70	2500
Polyester matrix	3.3	50

Problem 2.4 A tensile test is conducted on an aramid-epoxy sample 20 mm wide and 1 mm thick. The aramid composite has 50 vol% of unidirectional fibres in an epoxy matrix. If the tensile strength of the aramid fibres is 3600 MPa and the epoxy matrix is 50 GPa, estimate the axial failure load. Assume the stress in the matrix at fibre failure is 35 MPa.
Answer. 36.4 kN.

Problem 2.5 A UD glass-polyester composite has been selected for a structural tie in a truss. The mechanical properties of the E-glass and polyester are given in Table 2.7. It is believed that a fibre volume fraction of 0.35 can be consistently achieved during the manufacturing process. The tie is 50 mm wide and approximately 1 m long. If the composite tie is loaded to 70 kN and each ply is 0.34 mm thick, how many plies are needed to ensure

a. tensile failure does not occur?
b. the extension of the tie does not exceed 30 mm?

Note. Axial stiffness $F/\delta = AE/l$ where F is the tensile force, δ is the extension, A is the cross-sectional area, E is the elastic modulus and l is the length.
Answer. a. 5 plies; b. 6 plies.

Problem 2.6 A short rectangular-section beam is to be designed with the fixed width of 15 mm and a length of 0.2 m. The beam will be manufactured from a unidirectional carbon fibre-epoxy composite. The tensile modulus of the composite is 110 GPa, and the tensile and compressive strengths are 1750 MPa and 1000 MPa, respectively. If the beam is subjected to three-point bending test with a midspan load of 1 kN, determine the minimum number of 0.25 mm thick plies, to ensure

a. Longitudinal (tensile or compressive) failure does not occur.
b. The beam does not exhibit a midspan deflection of more than 3 mm.

Note. The second moment of area for a rectangular cross-section is $I = \frac{bh^3}{12}$ where b is the width and h is the height of the cross-section. Maximum deflection occurs at the midspan of the three-point bending test, and it is equal to $\delta_{max} = \frac{Fl^3}{48EI}$ where l is the length of the beam, and F is the point load.
Answer. a. 3 plies; b. 30 plies.

References

1. Callister WD, Rethwisch DG (2018) Materials science and engineering: an introduction, 10th edn. Wiley, Hoboken NJ
2. Mallick PK (2007). Fiber-reinforced composites: materials, manufacturing, and design, 3rd edn. Mechanical engineering. CRC/Taylor & Francis, Boca Raton
3. Wanberg J (2009) Composite materials: fabrication handbook #1, vol 1. Composite garage series. Wolfgang Publications, Stillwater, Minnesota
4. Barbero EJ (2017) Introduction to composite materials design, 3rd edn. Composite materials. CRC Press, Boca Raton
5. Hull D, Clyne TW (1996) An introduction to composite materials, 2nd edn. Cambridge solid state science series. Cambridge University Press, Cambridge
6. Adams D (2019) Optimum unidirectional compression testing of composites. www.compositesworld.com/articles/optimum-unidirectional-compression-testing-of-composites
7. Halpin JC, Tsai SW (1967) Environmental factors in composite design. air force materials laboratory
8. Javanbakht Z, Hall W, Virk AS, Summerscales J, Öchsner A (2020b) Finite element analysis of natural fiber composites using a self-updating model. J Compos Mater 54(23):3275–3286. https://doi.org/10.1177/0021998320912822
9. Javanbakht Z, Hall W, Öchsner A (2020a) An element-wise scheme to analyse local mechanical anisotropy in fibre-reinforced composites. Mater Sci Technol 36(11):1178–1190. https://doi.org/10.1080/02670836.2020.1762296
10. Voigt W (1889) Ueber die beziehung zwischen den beiden elasticitätsconstanten isotroper körper. Annalen der Physik 274(12):573–587. https://doi.org/10.1002/andp.18892741206
11. Reuss A (1929) Berechnung der fließgrenze von mischkristallen auf grund der plastizitätsbedingung für einkristalle. ZAMM - Zeitschrift für Angewandte Mathematik und Mechanik 9(1):49–58. https://doi.org/10.1002/zamm.19290090104
12. Hibbeler RC (2014) Statics and mechanics of materials, 4th edn. Pearson, Upper Saddle River N.J
13. Hart-Smith LJ (1992) The ten-percent rule for preliminary sizing of fibrous composite structures. Weight Eng 52:29–45
14. Kelly A, Tyson WR (1965) Tensile properties of fibre-reinforced metals: copper/tungsten and copper/molybdenum. J Mech Phys Solids 13(6):329–350. https://doi.org/10.1016/0022-5096(65)90035-9
15. Adams D (2017) Can flexure testing provide estimates of composite strength properties? www.compositesworld.com/articles/can-flexure-testing-provide-estimates-of-composite-strength-properties
16. Sun W, Guan Z, Li Z, Zhang M, Huang Y (2017) Compressive failure analysis of unidirectional carbon/epoxy composite based on micro-mechanical models. Chin J Aeronaut 30(6):1907–1918. https://doi.org/10.1016/j.cja.2017.10.002
17. DoITPoMS (2019) Strength of long fibre composites. www.doitpoms.ac.uk/tlplib/fibre_composites/strength.php
18. Medina C, Canales C, Arango C, Flores P (2014) The influence of carbon fabric weave on the in-plane shear mechanical performance of epoxy fiber-reinforced laminates. J Compos Mater 48(23):2871–2878. https://doi.org/10.1177/0021998313503026
19. Öchsner A (2016) Continuum damage and fracture mechanics, 1st edn. Springer, Singapore, Imprint: Springer Singapore
20. Clyne (2019) An introduction to composite materials. Cambridge University Press, Cambridge
21. Astrom BT (2018) Manufacturing of polymer composites, 2nd edn. Routledge, Boca Raton
22. Lokensgard E (2010) Industrial plastics: theory and application, 5th edn. Delmar Cengage Learning, Clifton Park NY
23. Krenchel H (1964) Fibre reinforcement; theoretical and practical investigations of the elasticity and strength of fibre-reinforced materials. Akademisk forlag
24. Lavender Composites (2017) Prepeg stock list. http://www.lavender-ce.com/wp-content/uploads/prepreg-ex-stock-101017.pdf

Chapter 3
How to Make a Composite—Wet Layup

Abstract This chapter introduces the six basic steps needed to design and make a fibre-reinforced composite (FRC): (1) fibre and matrix selection; (2) mould preparation; (3) layup and consolidation; (4) curing (and post-curing); (5) demoulding; and (6) post-processing (finishing). These six steps are considered using a simple wet layup process for a flat unidirectional (UD) composite. The mechanical performance of the composite is estimated based on the rule of mixtures (RoM), inverse rule of mixtures (IRoM) and Kelly-Tyson (KT) models from Chap. 2. A description of the wet layup process is offered and discussed in the context of the options available to a composite fabricator. The chemical crosslinking processes for polyester, vinylester and epoxy are presented. Consideration is given to demoulding the FRC and to the post-processing options. The use of gelcoat, flow coat and paint are mentioned in the context of a broader discussion on surface finishing of FRC. The outcome of the chapter is a step-by-step design and manufacturing method that is easily replicated.

Caution!
The manufacture of FRC structures demands the use of fibres and resins. The abrasive fibres, when cut or sanded, will result in fine dust particles that can cause skin irritations. Resins start as liquids and are mixed with either an initiator (catalyst) or hardener which then cures to create the solid polymer matrix that surrounds the fibres. The chemicals used in the curing process of thermosetting polymers can cause harm. Many other chemicals, including cleaning solvents and release agents, may also be used in the fabrication process [1]. For the composite manufacturing tasks covered in this text, it is the fabricator's sole responsibility to familiarise themselves with the associated risks for each of the activities.

Fig. 3.1 The six basic steps to make a FRC

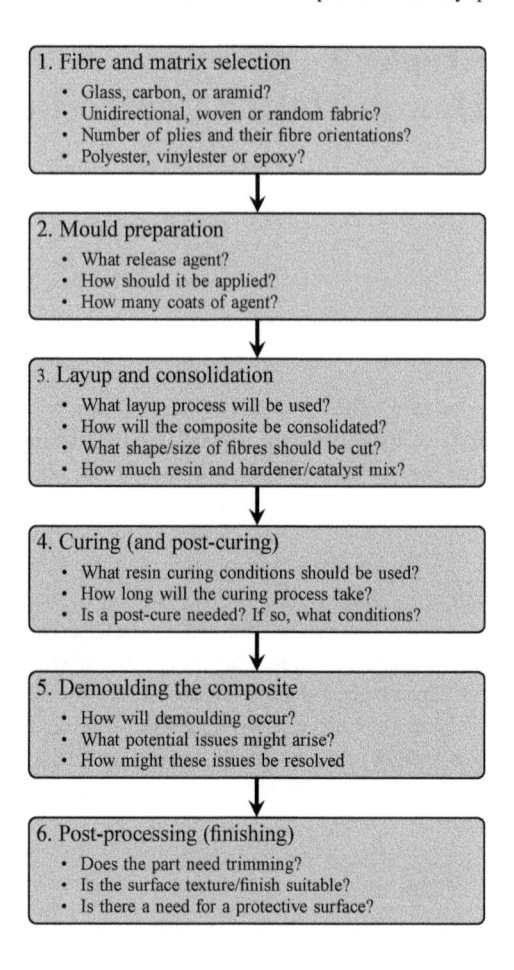

1. Fibre and matrix selection
   - Glass, carbon, or aramid?
   - Unidirectional, woven or random fabric?
   - Number of plies and their fibre orientations?
   - Polyester, vinylester or epoxy?

2. Mould preparation
   - What release agent?
   - How should it be applied?
   - How many coats of agent?

3. Layup and consolidation
   - What layup process will be used?
   - How will the composite be consolidated?
   - What shape/size of fibres should be cut?
   - How much resin and hardener/catalyst mix?

4. Curing (and post-curing)
   - What resin curing conditions should be used?
   - How long will the curing process take?
   - Is a post-cure needed? If so, what conditions?

5. Demoulding the composite
   - How will demoulding occur?
   - What potential issues might arise?
   - How might these issues be resolved

6. Post-processing (finishing)
   - Does the part need trimming?
   - Is the surface texture/finish suitable?
   - Is there a need for a protective surface?

3.1 Introduction

There are six basic steps to making a fibre-reinforced composite (FRC), as shown in Fig. 3.1. This step-by-step process is adapted from the six stages reported by Wanberg [2] and is considered in this chapter in the same logical and sequential manner. The sequence in Fig. 3.1 assumes that a mould for forming the composite is already available—design and construction of moulds for making composite parts is considered in Chap. 6. The six steps are common for basic manufacturing methods such as wet layup [2, 3] (also called hand lamination [4] or layup moulding [5]) as well as for more complex moulding techniques such as vacuum bagging [3, 5–7] and prepreg layup [3, 5] (referred to here as prepreg moulding [7, 8]). Wet layup, as the most basic method, is introduced in this chapter to provide context to the six-step process, whilst vacuum bagging and prepreg moulding methods are discussed and linked back to the same six steps in Chap. 4.

3.1 Introduction

Herein, the six steps are described for a flat UD carbon FRC panel. The manufacturing process for the FRC is used to facilitate a broader discussion of the manufacturing options available to the composite fabricator.

3.2 Fibre and Matrix Selection

To design a composite structure (or part), the applied loads and in-service conditions must be known. The reinforcing fibres and matrices are chosen to meet these structural (and environmental or in-service) requirements. The fibre and matrix selection step therefore relates to the 'Mechanics of Composite Structures' topic, covered earlier in Chap. 2.

To meet the structural requirements, the following considerations need to be addressed [9]:

- Fibre and matrix selections.
- Fibre volume fraction (linked to the manufacturing process).
- Fibre orientation in each layer or ply (and the stacking sequence).
- The number of plies in each orientation (and hence, the laminate thickness).

Here, to illustrate the wet layup process, UD carbon fibres (200 g/m²) [10] and an epoxy matrix [11] are chosen. The UD fabric is cut to approximately 0.3 m squared. Since wet layup is used, a relatively low fibre volume fraction is likely in comparison to other more advanced moulding methods since minimal consolidation pressure is provided during the wet layup process. An increase in consolidation pressure during moulding will reduce laminate thickness and hence, increase fibre volume fraction [12].

Typical fibre volume (and weight) fractions for wet layup (based on the fibre type and form) are shown in Table 3.1. Here, we assume $V_f \approx 0.4$: a mid-range value typical for wet layup of unidirectional composites. Four layers (plies) of carbon fibre are stacked and orientated in the same direction to create a UD FRC of approximately 1 mm thickness—see Eq. (5.5). These fibre and matrix selections are made to produce a composite that is stiff (has a high elastic modulus) and strong in the longitudinal or fibre direction, but is significantly more compliant in the transverse direction. Similar fibre and matrix selections will be used later in Chap. 4 for vacuum bagging and prepreg moulding to illustrate the comparative performance of wet layup against these other more advanced moulding methodologies (see Chap. 5).

Example 3.1

The wet layup method is used in this chapter to create a UD carbon fibre-reinforced epoxy laminate with a target fibre volume fraction $V_f \approx 0.4$. Use Eqs. (2.1)–(2.6) to estimate the elastic moduli and the tensile strengths (longitudinal and transverse) of the FRC. The fibres and resin are identified in Refs. [10, 11].

Solution

Wet layup is a simple manufacturing process but is likely to produce relatively low fibre volume fractions (0.3–0.5) and hence, we assume that $V_f \approx 0.4$ (see Table 3.1).

The elastic moduli and strength of the fibres are reported to be 250 GPa and 5516 MPa, respectively [10]. The tensile modulus and strength of the epoxy resin are not readily available; *typical* values have been estimated, based on mechanical properties for similar commercial resin systems. The elastic modulus and tensile strength are assumed here to be 3.0 GPa and 50 MPa, respectively.

Considering $E_m = 3.0$ GPa and $E_f = 250$ GPa, the longitudinal modulus is calculated from

$$E_{cl} = E_m(1 - V_f) + E_f V_f,$$
$$= 101.8 \times 10^9 \, \frac{N}{m^2} = 101.8 \, \text{GPa}$$

The transverse modulus is calculated from

$$\frac{1}{E_{ct}} = \frac{(1 - V_f)}{E_m} + \frac{V_f}{E_f},$$
$$E_{ct} = 5.0 \times 10^9 \, \frac{N}{m^2} = 5.0 \, \text{GPa}$$

Assuming $\sigma_f = 5516$ MPa, the longitudinal strength is calculated from

$$\sigma_{cl}^* = \sigma_{m(f^*)}(1 - V_f) + \sigma_f^* V_f$$

or conservatively, neglecting the matrix contribution

$$\sigma_{cl}^* = \sigma_f^* V_f$$
$$= 2206.4 \times 10^6 \, \frac{N}{m^2} = 2206.4 \, \text{MPa}$$

The transverse strength is calculated from

$$\sigma_{ct}^* \ll \sigma_m^*$$
$$\sigma_{ct}^* \ll 50 \, \text{MPa}$$

Note. Later (in Chap. 5), we find the measured V_f for the panel is actually 0.34, rather than the targetted value of 0.4.

3.3 Mould Preparation

Table 3.1 Fibre fractions for wet layup (hand lamination) based on fibre type and form

	Random mat		Woven		Unidirectional	
	V_f	W_f	V_f	W_f	V_f	W_f
Glass	0.1–0.3	0.19–0.48	0.2–0.4	0.35–0.59	0.3–0.5	0.48–0.68
Carbon	0.1–0.3	0.14–0.38	0.2–0.4	0.27–0.49	0.3–0.5	0.38–0.59
Aramid	0.1–0.3	0.12–0.34	0.2–0.4	0.23–0.44	0.3–0.5	0.34–0.55

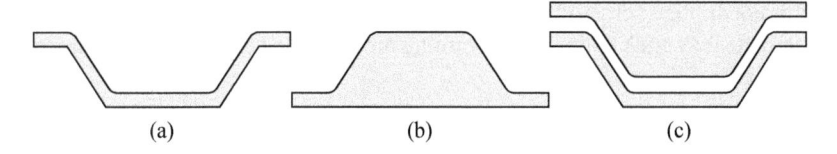

Fig. 3.2 Mould shapes and terminologies: **a** female; **b** male; and **c** matched-die mould

3.3 Mould Preparation

A FRC is usually formed in a mould, often referred to as *tooling* [3, 5, 13]. A mould can be manufactured from a wide variety of materials. The design of the mould depends on the composite structure to be created and on the number of parts to be produced by the mould tool (production run). Further details about mould design, tooling material options, how to create mould tools and other basic moulding considerations are provided in Chap. 6.

As a starting point, let's assume our mould tool is a flat plate. A flat plate will produce a composite laminate with a single smooth surface on the mould side. The fibres and surrounding matrix will be pressed against the flat mould surface. On the reverse side of the composite, a less appealing surface finish will be produced as the fibres are not pressed firmly against a surface. More complex mould tool shapes can produce concave or convex surfaces, but still with a single smooth surface [13, 14].

A *female* mould is identified by its concave form and is used to create a (convex) laminate with a smooth outer surface, whilst a *male* mould has a convex shape and is used to create a smooth inner (concave) surface on the composite [3, 7, 14]. To create a smooth surface on both sides normally requires a *matched* mould [7] or matched-die moulding system [3, 14], i.e. a combination of both male and female mould tools. However, it is also possible to simulate a two-sided mould with a simple caul plate [14, 15]. A caul plate is a smooth metal or plastic plate (or sometimes a rubber sheet) used to provide a uniform normal pressure to the laminate during moulding and hence, a smooth surface on the 'reverse' side of the cured composite part. Female, male and matched-die moulds are shown in Fig. 3.2. The aesthetic requirements of the composite therefore play a significant role in the selection of a suitable moulding system.

An appropriate mould preparation process is essential for the successful creation of moulded composite structures. Mould tools should be clean and *released* prior to commencement of the layup process (described in Sect. 3.4) [2, 16]. To remove contaminants, a clean cloth (ideally, a lint-free cloth) and a solvent (for example, acetone) are typically used. The mould tool should then be left to dry before the *release agent* [17] (sometimes called parting agent [13]) is applied to the moulding surface. The release agent is used to provide a thin uniform barrier between the mould tool and the composite laminate, preventing permanent adhesion [18]. If a release agent is not applied across the entire mould surface (or is not appropriately applied), the laminate may stick to the mould, ruining the laminate and/or rendering the mould tool unusable.

A laminate is more likely to stick to a new mould tool than a well- used one [19]. Thus, new moulds tend to require additional preparation whereas older (sometimes referred to as *seasoned* [2]) moulds need fewer release coats. As a rule of thumb, 2–3 coats of release agent are usually needed on a seasoned mould, but a minimum of 4 is suggested in [2] for new tooling. There are many types of release agents [20] but three types of mould release that are readily available from most composite suppliers are:

- Wax.
- Polyvinyl alcohol (PVA).
- Semi-permanent release agents.

In this text, wax is exclusively used as the release agent. Standard wax mould release is normally used for low-temperature curing processes (room-temperature cures) but some high-temperature wax formulations are used later for elevated temperature cures up to 120 °C—see Chap. 4. In Fig. 3.3, standard mould release wax is applied to the mould tool (flat glass plate). Masking tape is applied around the plate to define the region of wax application. When applying the wax (or any other release agent), careful consideration should be given to specific application instructions. Here, three coats of wax are applied to the seasoned flat plate with circular motions. A small quantity of wax is applied over the entire mould with a sponge. The wax is allowed to dry (a haze is observed on the surface) and then the surface is buffed to a sheen with a clean lint-free cloth. The surface is left for a further time period before the next coat of wax is introduced—the requisite timings are specified in the product's instructions.

Polyvinyl alcohol and semi-permanent release agents are not used in this textbook, but their widespread availability and popularity amongst many composite fabricators means they (at the very least) warrant a brief comment—wax is used consistently here solely as personal preference. Polyvinyl alcohol can be used on its own as a sole release agent, or in some circumstances as a secondary mould release for composite parts where demoulding is problematic. If needed, and depending on the PVA product chosen, PVA can be wiped, brushed or sprayed on the mould surface [21].

Semi-permanent release agents are another alternative to wax that can help speed up the release agent application process. Sometimes, these releases are referred to as *polymer release agents* since they comprise one or more polymer resins dissolved

Fig. 3.3 Application of a wax mould release

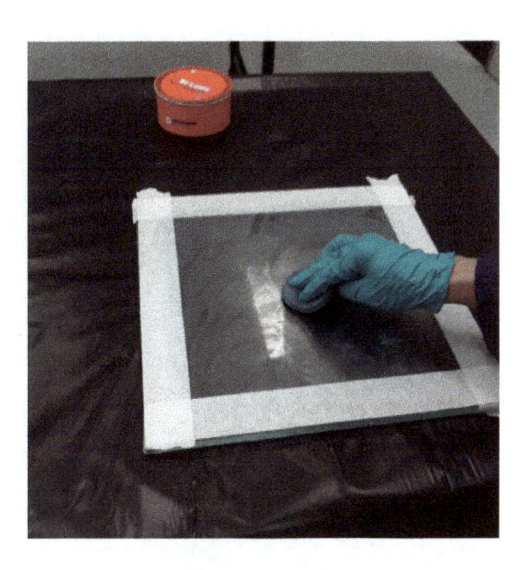

in a solvent [17]. Semi-permanent release agents can offer multiple moulding cycles before reapplication of the release agent is necessary. If PVA or a semi-permanent releases are applied to a mould tool, they should always be allowed to dry before the composite layup process commences.

3.4 Layup and Consolidation (Hand Lamination)

The first surface layer of a laminate is an optional gelcoat (a layer of thickened resin); typically, a gelcoat is 0.4–0.6 mm thick [22] but may be applied to a thickness of up to 1 mm [23] or more. A gelcoat, whether polyester-, vinylester- or epoxy-based, is used to provide a high-quality finish on the surface of the composite and to protect the laminate from environmental factors. Most often, gelcoats are polyester- or vinylester-based and tend to contain a pigment to provide a smooth coloured aesthetic appearance [5]. The hard (often white) outer surface of a boat hull is an example of a gelcoat finish on a glass fibre composite.

Gelcoats can be purchased directly from composite suppliers but if you wish to formulate your own, resin thickeners (gelcoat additives) and pigments can in most cases be added to standard lamination resins [2]. Where a gelcoat is used, a thin layer is brushed or sprayed onto the mould tool surface. If a gelcoat is not applied, either a clear flow (top) coat or paint [3, 24, 25] would normally be used to produce the finished look and (perhaps most critically) provide environmental protection for the composites. A flow coat or paint is applied as the last finishing process. Here, no gelcoat or finish coats (neither flow coat or paint) are used, but finishing applications on composite parts will be illustrated later in this textbook. Further information in relation to surface preparation for flow coat or painted surface finishes is given in Sect. 3.8.

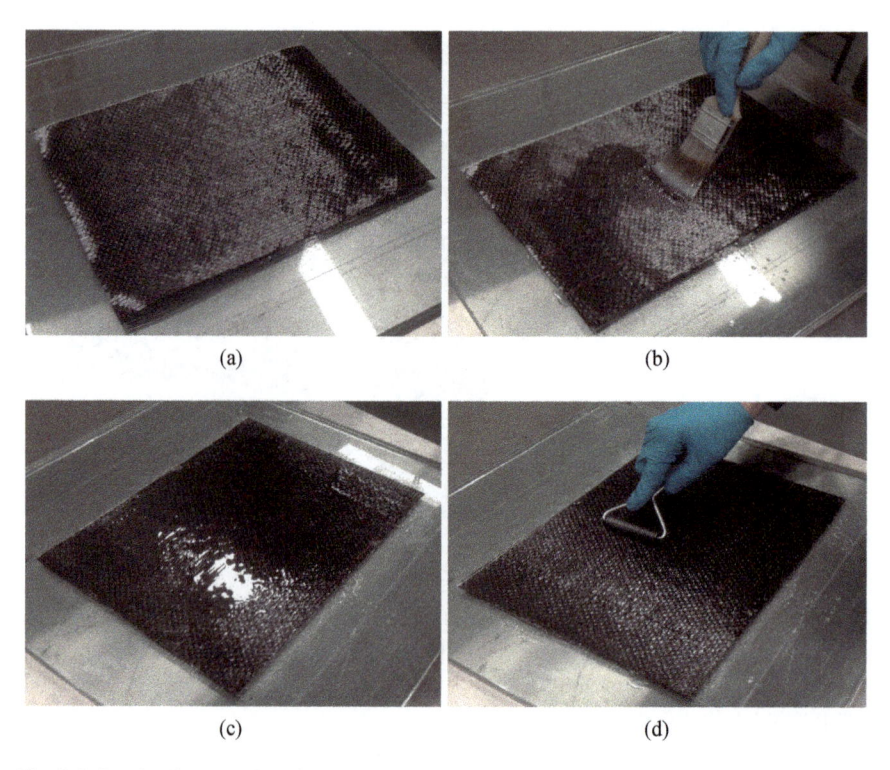

Fig. 3.4 Lamination: **a** carbon fibre layup (first ply); **b** wetting out with a brush; **c** excess surface resin; and **d** application of roller to remove air trapped between the plies

As no gelcoat is used here, the first carbon fibre ply is positioned on the flat plate. The resin is mixed. During the mixing process, additional substances such as pigments may be added [26] without an adverse effect on the resin characteristics (or finished composite). The premixed resin (in our case, epoxy [11]) is introduced to the carbon fibre; an excess of resin is sometimes added to the first ply to ensure resin adequately impregnates the fabric. A brush is often used (a spreader is an alternative option) to ensure a uniform distribution of resin across the surface of each of the carbon fibre plies. The manual fibre positioning and the introduction of the resin are shown in Fig. 3.4a and b, respectively. In some cases, an alternative wet-out process is employed; a small quantity of resin is applied to the mould surface prior to the placement of the first ply—this is an alternative method to ensure a smooth surface finish on the mould side of the composite. The wetting-out process is repeated for each ply. After each layer is wet out, a roller is used to assist resin impregnation and expel trapped air —see Fig. 3.4d. A roller will have minimal effect on consolidation.

Wet layup is relatively easy to perform, but to achieve the best results, it does require some practise. It is easy to use too much resin and add unnecessary mass to the composite. *So, how much resin is needed?*

3.4 Layup and Consolidation (Hand Lamination)

The mass of the resin (and catalyst/hardener mix) can be calculated from the area of the laminate (A), the number of layers (n), the areal weight of the fibres (A_w) and the intended weight fraction of the matrix (W_m) [27]

$$\text{Resin and hardener mix (g)} = \frac{A \times n \times A_w \times W_m}{(1 - W_m)} \quad (3.1)$$

or in terms of fibre weight fraction

$$\text{Resin and hardener mix (g)} = \frac{A \times n \times A_w \times (1 - W_f)}{W_f} \quad (3.2)$$

These equations offer simple estimates of the mass of the resin and hardener mix needed to create the laminate, but they do not account for any resin wastage that may occur during fabrication. A factor between 1 (assuming no waste) and 1.5 is therefore usually introduced to take account of resin wastage [27]. The factor is usually chosen with an understanding of the composite layup and is based on the fabricator's expertise.

> **Example 3.2**
>
> Now, using Eq. (3.2), calculate how much resin and how much hardener are needed to fabricate the UD carbon fibre-reinforced epoxy laminate shown in Fig. 3.4.

> **Solution**
>
> As mentioned, the fibres were cut to 300 mm long × 300 mm wide. The surface area of the fibre-reinforced panel was therefore (0.3 m squared) = 0.09 m².
>
> The fibre volume fraction was previously chosen based on the unidirectional carbon fibre values for wet layup in Table 3.2, i.e. $V_f = 0.4$ which is at the midpoint between the upper and lower threshold values.
>
> Thus, $W_f = 0.49$ and hence
>
> $$\text{Resin and hardener mix (g)} = \frac{A \times n \times A_w \times (1 - W_f)}{W_f}$$
>
> $$\text{Resin and hardener mix (g)} = \frac{0.09 \times 4 \times 200 \times (1 - 0.49)}{0.49} = 75\,g$$
>
> To account for wastage, a factor of 1.25 (between 1 and 1.5) is used for the resin and hardener mix.

$$\text{Resin and hardener mix (g)} = 75 \times 1.25 = 94\,\text{g}$$

The epoxy resin used here has a resin to hardener ratio of 5:1, i.e. $^5/_6$ of resin and $^1/_6$ of hardener

$$\text{Resin (g)} = 94 \times {}^5/_6 = 78.3\,\text{g}$$
$$\text{Hardener (g)} = 94 \times {}^1/_6 = 15.7\,\text{g}$$

Note. An appropriate selection of the wastage factor is an important skill for a composite fabricator but it is simply a sensible estimate (based on experience) for the composite being fabricated. Adding extra resin will, of course, reduce V_f whilst not enough will result in a dry composite with suboptimal properties.

3.5 Curing and Post-Curing

Once the layup process is complete, the panel is left to cure. There will be chemical and structural changes during the curing process, and the resin will transform from a low molecular weight liquid into a solid polymer [28]. The cure time depends on many factors but most need a curing time of at least 24 h at room temperature [29] to reach the solid state.

During curing, heat will be discharged (referred to as an *exothermic* reaction [30]) and the viscosity of the resin will increase. The resin starts to *gel*—this process typically takes 30–40 min (sometimes more, sometimes less) and is dependent on the type and volume of the resin and the ambient temperature. As a rule of thumb, a 10 °C temperature increase will half the working time [27]. Moreover, it should be noted that a thick laminate can generate a considerable amount of heat and could be a potential safety hazard, whilst minimal heat will usually result from a thinner laminate; the peak exothermic temperature increases with thickness [5], so care should be taken in the manufacture of thicker composite structures.

After the tacky, gelatinous stage, the composite will continue to cure and gradually become firm (sometimes referred to as the *green* stage [2]) before finally reaching the solid state. Once the solid state is attained, the laminate has sufficient stiffness to be demoulded (see Sect. 3.7), but further crosslinking may still be needed. Epoxy resins often benefit from a post-cure at elevated temperatures to reach their optimum performance values [31]—see Ref. [27]. Here, a post-cure is performed after demoulding. Post-curing can be performed before or after demoulding [32], but is usually performed after removal from the mould [29]. A post-cure should always be performed in accordance with the supplier's instructions [5, 33].

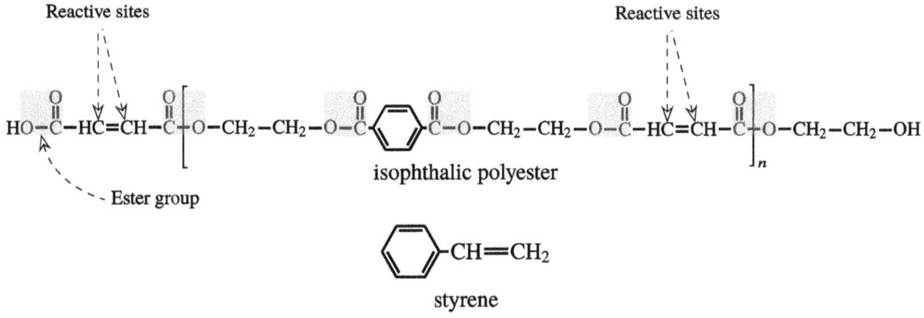

3.6 Crosslinking: Chemical Steps

Fig. 3.5 The chemical structure of an isophthalic polyester and styrene (ester groups are highlighted). Adapted from [5]

3.6 Crosslinking: Chemical Steps

As mentioned, during curing, the liquid polymer is irreversibly transformed into a solid, but what chemical changes occur during this process? The details of the chemical crosslinking process for polyester, vinylester and epoxy are elaborated in this section.[1]

Unsaturated Polyester. The production of polyester is usually the result of reacting unsaturated and saturated acids and a glycol (an alcohol) to produce a short chain polymer, which is usually a stable liquid. Adjusting the amount of unsaturated acid allows for the creation of potential sites for crosslinking whereas the amount of saturated acid and glycol determines the chemical and fire resistance [33].

The chemical structure of a typical unsaturated (isophthalic) polyester is shown in Fig. 3.5. The intermolecular bonds of the polyester are represented as a line between two atoms, i.e. a pair of shared electrons. These bonds are covalent and the term *unsaturated* identifies the existence of carbon double bonds in the backbone of the chain [5]. The letter 'n' identifies the part of the molecule (in square brackets) that is repeated n-times (typically $n = 3$ to 6 [27]). The double bonds are the reactive sites that provide a *bridge* between polyester chains during the crosslinking process. A small molecule such as styrene (a volatile organic compound, VOC) is used as a reactive diluent [5] to enhance the reactivity of the liquid polymer (and to reduce cost) whilst also facilitating the crosslinking process without the evolution of by-products. The styrene molecule shown in Fig. 3.5 acts as the bridge between the polyester chains.

A simplified representation of the molecular chains during the curing process is presented in Fig. 3.6. The polyester chains are shown as a simple 2D structure but will (in reality) form a 3D arrangement. The curing process starts when all the ingredients are available: the unsaturated polyester (containing the styrene diluent)

---

[1] The term *crosslinking* should not be confused with *polymerisation*; the latter happens during manufacturing of a polymer from monomers, whilst the former occurs during the moulding (or curing) process when polymer chains join together [34].

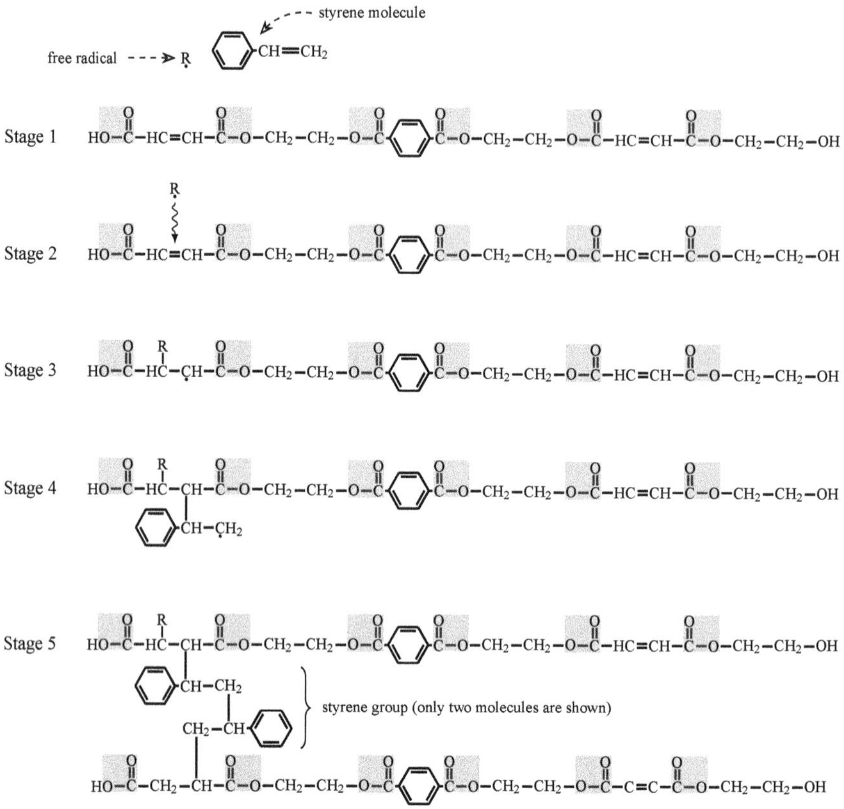

Fig. 3.6 Simplified representation of crosslinking in an unsaturated polyester (ester groups are highlighted). Adapted from [5]

and an initiator (typically referred to as a catalyst[2]). An accelerator is already included in the unsaturated polyester to promote the crosslinking reactions.

The catalyst is usually peroxide-based and the most common is methyl ethyl ketone peroxide (MEKP) [5]. The accelerator decomposes the initiator/catalyst into *free radicals* (i.e. molecules with unpaired electrons) [33], which are denoted by Ṙ in Fig. 3.6 (Stage 1). These free radicals seek to react with loosely bonded electrons in the carbon-carbon double bond (Stage 2). When a free radical reacts with an electron from the double bond, the carbon-carbon double bond is broken and the free radical forms a single bond with carbon. This leaves the carbon chain with a single carbon-carbon bond and one unpaired electron (Stage 3). The unsaturated carbon-carbon bond of a styrene molecule bonds with this unpaired electron (breaking its double bond) and leaving an unpaired electron (Stage 4). As this process repeats,

---

[2] The initiator is often referred to as the catalyst but, since its structure changes during the reaction, this is strictly not accurate; the accelerators are technically the catalysts in the current context [33]. The term *catalyst* is employed in this text as this is the name typically used in the composite industry.

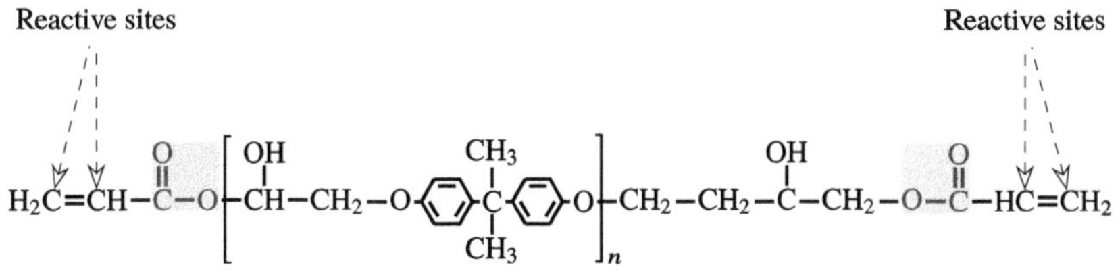

3.6 Crosslinking: Chemical Steps

Fig. 3.7 The chemical structure of a typical vinylester (with an epoxy backbone). Adapted from [9, 27]

Fig. 3.8 The chemical structure of a typical DGEBA epoxy. Adapted from [5, 9, 27]

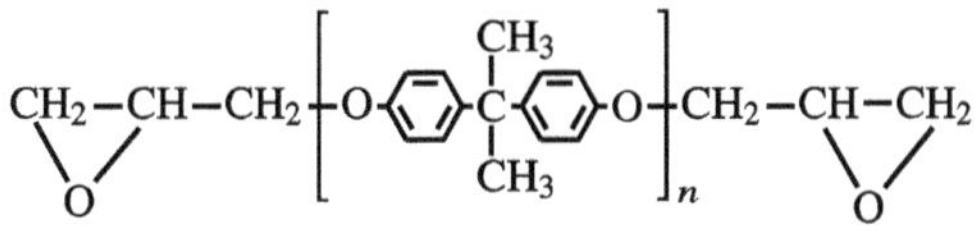

the crosslinking bridge continues to grow until another polyester change comes into the reaction to terminate the process; see Stage 5 in Fig. 3.6. Note that in the same figure, two styrene molecules are represented schematically as the bridge between the polyester chains, but evidently this styrene group could contain only a few or many styrene molecules. The final result is several polyester main chains with crosslinking styrene groups between them [9].

Vinylester. A typical vinylester molecule is shown in Fig. 3.7. The vinylester molecule has an epoxy backbone with ester linkages and double bonds (reactive sites) at each end instead of epoxy rings (cf. Figs. 3.7 and 3.8) [5]. The crosslinking process for vinylester is similar to that for polyester and commences with a catalyst, but there are fewer reactive sites. This chemical structure helps to make vinylester tougher (less brittle) than its polyester counterpart [27]. Increasing the number of styrene molecules in the crosslinks can cause the cured polymer to exhibit the brittle characteristics of polystyrene. Moreover, having an epoxy backbone along with a reduction in ester groups makes vinylester less susceptible to hydrolysis compared to unsaturated polyesters [5, 6].

Epoxy. The chemical structure for a typical epoxy is shown in Fig. 3.8. Epoxies are characterised by three-member rings (reactive sites) containing two carbon atoms bonded to an oxygen atom [9]. Whilst epoxy rings can be located either at the extremities or in the middle of the polymer chain [14], for Diglycidyl Ether of Bisphenol-A (DGEBA), these reactive sites are at either end of the epoxy molecule. As mentioned earlier, unlike polyester and vinylester that cure by a catalyst, epoxies cure with a hardener. The hardener is usually an amine curing agent containing reactive NH_2 groups to facilitate 'opening' of the epoxy ring [5]. There are usually two reactive groups at the end of the curing agent to allow a reaction with two epoxy molecules.

A simplified representation of the crosslinking process of an epoxy ring is given in Fig. 3.9. The reaction commences when the epoxy and the hardener are mixed, i.e. when the reactive NH_2 group encounters the epoxy ring. Then, a C–O bond in the

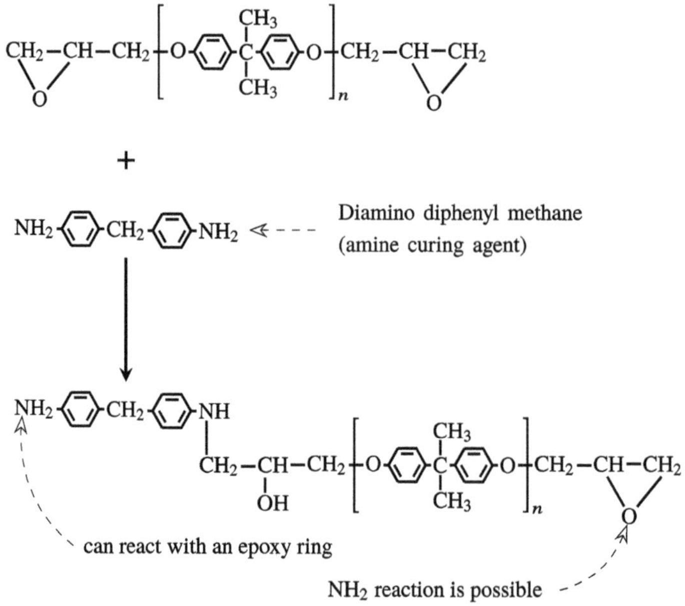

Fig. 3.9 Simplified representation of epoxy crosslinking with an amine hardener. Adapted from [9]

epoxy and an N–H bond in the hardener are broken and alternative bonds between the epoxy and hardener are created, i.e. a C–N and a C–OH bond. The crosslinking process continues as the remaining reactive site, at the other end of the hardener molecule, can react with another epoxy molecule, which results in connecting two epoxy chains together [9]. The epoxy and amine molecules co-react, and hence require a specific ratio. If they are not mixed in the correct proportions, unreacted resin or hardener will result, and this will ultimately reduce the cured polymer properties [5]. Note that the lack of ester groups ensures epoxies are less susceptible to moisture uptake compared to unsaturated polyester or vinylester [5, 6, 27].

3.7 Demoulding

After a laminate is cured to the solid state, it can be demoulded. Here, to remove the FRC (panel) from the flat plate, a soft plastic wedge is used—see Fig. 3.10 . For laminates with more complex shapes, some extra encouragement may be needed. Assisted demoulding processes that involve mechanical [2], temperature [35] or compressed air extraction methods [29] will be discussed in Chap. 7. The purpose of the wedge in our case is to simply lift one corner of the panel from the flat plate without causing damage to either the UD composite or mould. Demoulding should be relatively easy to achieve for a smooth flat plate mould that has been appropriately released. Once a small part of the FRC is freed from the mould, the rest of the panel

3.7 Demoulding

Fig. 3.10 Demoulding with a plastic wedge

should release with relative ease. Note that care should always be taken during the demoulding process as it is easy to damage a laminate or, in some cases, the mould. It is *never* advisable to use metal hand tools (or any other hard or abrasive materials) for demoulding as these may scratch or score the mould tool.

3.8 Post-Processing (Finishing)

A FRC will rarely leave the mould in the finished state [3]. As a minimum, trimming and sanding are usually required. Moreover, laminates are often assembled (or joined) to other traditional materials (metals, ceramics and polymers) or other composite parts to form larger assembled structures [9]. These joining processes often involve drilling [36], cutting or machining operations [37]; even bonding composites to other materials requires some abrasion of the surface [5, 15]. The post- (or secondary) processing activities should be tackled with caution as these actions can cause damage to a laminate (as well as rapid tool wear in the case of machining fibrous composites) [38].

Machining Processes. Trimming, cutting, drilling and other machining processes can often be performed (with care) using basic hand tools or conventional metal working equipment, e.g. lathes and milling machines [39]. In general, the cutting speeds for composite materials should be higher and the feed rates lower than those selected for similar operations on metals [3, 5]. This minimises the forces and temperature effects [40] that can cause damage. In the same context, the use of carbide (medium cost), diamond or boron nitride (high cost) tooling is normally recommended for FRCs as traditional cutting tools can blunt quickly since fibre reinforcements are abrasive [41]. The use of blunt cutting tools increases tool friction and hence thermal issues.

To finish the demoulded carbon fibre panel in Fig. 6.16, it is marked to the appropriate dimensions and then simply and carefully trimmed using a table saw, but a hacksaw or bandsaw are suitable alternatives. If a more complex shape is needed, a jigsaw can be used to cut the outline from the panel. The optional gelcoat is not integrated into our composite panel, and no extra surface coating is added here at this post-processing phase; a surface coating is normally applied to a composite to provide environmental protection and a visual appearance, but neither are considered important in this instance. The panel will be used later to benchmark the mechanical properties attained from wet layup in comparison to other advanced manufacturing methods, and to benchmark the performance of the RoM and KT equations—see Chap. 2.

Surface Finish. If a surface finish is needed at the post-processing step, options include the use of a flow coat or paint (clear and coloured paints). A flow coat is a viscous coating of resin that covers the external surface of the composite. The application of both a flow coat and painting requires similar initial surface preparation methods.

In terms of preparation for a flow coat or paint, the surface is cleaned to remove debris or contaminates (including residual mould release) and then sanded to provide a smooth finish. A clean cloth and warm water, and/or an appropriate solvent are used to clean the surface. Sanding helps remove tiny imperfections from the surface of the laminate but should ideally only be used to roughen the surface and not expose the fibres [3]; to prevent dust, wet sanding is usually recommended. Sometimes surface filling is also necessary, but this is less common when a composite is made using a mould. Sanding often commences at 120–220 grit sandpaper (<P120–P220) to initially remove the high spots and then moves through a series of grades until finishing is completed with a fine 400 grit paper (P600–P800) or more—Wanberg [2] suggest 320–600 grit. Once sanding is complete, the surface should be cleaned and left until completely dry.

The flow coat or paint layers are applied after drying. Paints can often allow the addition of a primer for maximum adhesion between the laminate surface and the top coat [42]. Primer fillers may also need to be used to remove any remaining imperfection, prior to application of the finish coat. Following the supplier's recommendations, additional sanding at very fine grits ($\gg$400 grit) is sometimes needed between each coat.

Example 3.3

Consider the wet layup process. List what you think are the main advantages and disadvantages.

3.8 Post-Processing (Finishing)

Solution

The main advantages and disadvantages of wet layup process are summarised as follows [25, 27, 43]:

Advantages	Disadvantages
Simple and versatile method. The wet layup process is simple to understand and can be used in many applications—from small parts to large composite structures.	*Fibre orientation and resin mixing.* The fibre orientation is aligned by the fabricator and resin mixing is performed by hand. Thus, laminate quality depends on the fabricator.
Low-cost manufacturing process. Minimal equipment is needed to produce a fibre-reinforced composite and hence, it is cost-effective for prototyping or small production runs.	*Fibre volume fraction.* Lower fibre volume fractions (and higher void contents) are attained in comparison to other advanced methods since minimal consolidation can be achieved via a roller—see Chap. 4.
	Laminate consistency. Maintaining consistency in a composite laminate and between samples is difficult, and hence, a greater variation can often be seen in laminate performance.
	Health and safety considerations. The resin mixing process is a manual process, and fabricators are exposed to volatile organic compounds (VOCs).

3.9 Summary

There are six basic steps to make a FRC:

1. Fibre and matrix selection.
2. Mould preparation.
3. Layup and consolidation.
4. Curing (and post-curing).
5. Demoulding.
6. Post-processing (finishing).

The wet layup process is a simple, low-cost and versatile composite manufacturing process. Fibres are impregnated with resin manually and allowed to cure, usually at room temperature.

Table 3.2 Weight fractions for wet layup (hand lamination) based on fibre type and form

Fibre type	Random mat W_f	Woven W_f	Unidirectional W_f
Glass	0.19–0.48	0.35–0.59	0.48–0.68
Carbon	0.14–0.38	0.27–0.49	0.38–0.59
Aramid	0.12–0.34	0.23–0.44	0.34–0.55

Resin and hardener (or catalyst) mix can be calculated in terms of the weight fractions from

$$\text{Resin and hardener mix [g]} = \frac{A \times n \times A_w \times (1 - W_f)}{W_f}$$

3.10 Questions

Question 3.1 State the six basic steps needed to design and manufacture a FRC.

Question 3.2 Why is it important to clean and release a mould prior to commencing a composite layup process?

Question 3.3 What is a *gelcoat* and why would you use one?

Question 3.4 What is a *flow coat* and how does it differ from a gelcoat?

Question 3.5 Describe a simple wet layup and consolidation process.

Question 3.6 What fibre volume fractions are likely to be achieved for each of the following fibre reinforcements when using wet layup as a manufacturing process?

a. Plain weave carbon.
b. Unidirectional aramid.
c. Random (chopped strand) glass mat.

Question 3.7 Describe what is meant by *curing* and *post-curing*.

Question 3.8 After mixing, how long does it typically take for a resin to start to *gel*? How long should it take to transform from a low molecular weight liquid to a solid polymer?

Question 3.9 Describe the series of chemical steps involved in the curing (or crosslinking) reaction of an unsaturated polyester mixed with a small amount of initiator/catalyst. In doing so, identify the most common catalyst and reactive diluent used in the crosslinking reaction.

Question 3.10 Describe the series of chemical steps involved in the curing (or crosslinking) reaction of an epoxy mixed with an amine hardener.

3.10 Questions

Question 3.11 Why should you never use metal hand tools (for example, a scrapper) when demoulding a composite part?

Question 3.12 A FRC laminate will rarely leave the mould in the finished state. What post-processing operations are commonly required?

Question 3.13 What surface preparation is needed before a FRC is painted?

3.11 Problems

Problem 3.1 A FRC panel (0.2 m squared) is to be manufactured using wet layup. The laminate will comprise six layers of woven glass fibres (areal weight = 300 g/m²) in a polyester matrix. The resin specifies 2% MEKP catalyst by weight:

a. What fibre volume and weight fractions would you suggest?
b. What mass of the resin and catalyst mix should be used?
c. What mass of polyester resin and MEKP catalyst are needed?

Answer. a. $V_f = 0.3$, $W_f = 0.47$; b. 101.5 g (waste factor = 1.25); c. 99.5 g resin, 2.0 g catalyst.

Problem 3.2 How much polyester resin and catalyst mix are required to wet out a single layer (1 m²) of E-glass random mat with an areal weight = 450 g/m²?
Answer. 1918.4 g ($W_f = 0.19$)–487.5 g ($W_f = 0.48$).

Problem 3.3 Wet layup is used to manufacture a random FRC consisting of 20 vol% glass fibres in a polyester matrix. Assume E-glass fibres have a modulus of 70 GPa and the polyester has a modulus of 2.9 GPa. Estimate the composite's in-plane elastic modulus.
Answer. $E_c = 16.3$ GPa.

Problem 3.4 A unidirectional carbon fibre-reinforced epoxy composite is produced by wet layup. The tensile strength of the carbon fibres and the epoxy matrix is 3800 MPa and 70 MPa, respectively. If the FRC is axially loaded in tension (parallel to the fibres) until failure occurs, estimate its failure strength.
Hint. You will need to assume a fibre volume fraction appropriate for wet layup.
Answer. $\sigma_{cl}^* = 1520$ MPa (assuming $V_f = 0.4$).

References

1. Composites Australia (2019) Mould release agents. https://www.compositesaustralia.com.au/for-industry/health-and-safety/mould-release-agents/
2. Wanberg J (2009) Composite materials: fabrication handbook #1, vol 1. Composite garage series. Wolfgang Publications, Stillwater, Minnesota

3. Astrom BT (2018) Manufacturing of polymer composites, 2nd edn. Routledge, Boca Raton
4. Thomas S (2014) Polymer composites. Wiley-VCH, Weinheim
5. Strong AB (2008) Fundamentals of composites manufacturing: materials, methods and applications, 2nd edn. Society of Manufacturing Engineers, Dearborn, Mich
6. Barbero EJ (2017) Introduction to composite materials design, 3rd edn. Composite materials. CRC Press, Boca Raton
7. Wanberg J (2010) Composite materials: fabrication handbook #2. Composite garage series. Wolfgang Publications, Stillwater, Minnesota
8. Mayer RM (1993) Design with reinforced plastics: a guide for engineers and designers. Springer, Netherlands, Dordrecht. https://doi.org/10.1007/978-94-011-2210-8
9. Mallick PK (2007) Fiber-reinforced composites: materials, manufacturing, and design, 3rd edn. Mechanical engineering. CRC/Taylor & Francis, Boca Raton
10. Hyosung Corporation (2017) Tansome carbon fiber. https://www.hyosungusa.com/files/advanced/tansome_catalog_2017.pdf
11. Fiber Glass International (2021) Es180 epoxy data system. https://martglass.com.au/wp-content/uploads/FGI-Epoxy-Resin.pdf
12. Quinn J (ed) (1990) Compliance of composite reinforcement materials
13. Lee SM (1989-) Reference book for composites technology. Technomic Pub. Co, Lancaster Pa. U.S.A
14. Wang R, Zheng S, Zheng Y (2011) Polymer matrix composites and technology. Woodhead publishing in materials, Woodhead Pub. and Science Press, Oxford and Philadelphia and Beijing
15. Campbell FC (2004) Manfacturing processes for advanced composites. Elsevier Advanced Technology, Oxford
16. Akovali G (2001) Handbook of composite fabrication. Woodhead Publishing, Shawbury
17. Composites Australia (2019) Health and safety. https://www.compositesaustralia.com.au/for-industry/health-and-safety/
18. Clark SL (2013) Release agents. In: Lubin G (ed) Handbook of composites. Springer, New York, NY, pp 633–638
19. Potter K (1997) An introduction to composite products: design, development and manufacture. Chapman & Hall, London
20. Hyer MW (2009) Stress analysis of fiber-reinforced composite materials, updated. DEStech Publications Inc, Lancaster, Pennsylvania
21. Steele GL (1962) Fiber glass: projects and procedures. McKnight & McKnight
22. Yuhazri MY, Haeryip S, Zaimi ZA et al (2015) A review on gelcoat used in laminated composite structure. Int J Res Eng Technol 4:49–58
23. Bunsell AR, Renard J (2005) Fundamentals of fibre reinforced composite materials. Series in materials science and engineering, Institute of Physics Publishing, Bristol
24. Owen MJ, Middleton V, Jones IA (2000) Integrated design and manufacture using fibre-reinforced composites. Woodhead, Cambridge
25. Karlsson KF, TomasÅström B (1997) Manufacturing and applications of structural sandwich components. Compos Part A: Appl Sci Manuf 28(2):97–111. https://doi.org/10.1016/S1359-835X(96)00098-X
26. Fan M, Fu F (2016) Advanced high strength natural fibre composites in construction. Woodhead Publishing, Oxford
27. Gurit (2019) Guide to composites. https://www.gurit.com/Our-Business/Composite-Materials
28. Vergnaud JW, Bouzon J (1992) Cure of thermosetting resins: modelling and experiments. Springer, London. https://doi.org/10.1007/978-1-4471-1915-9
29. Weatherhead RG (1980) FRP technology: fibre reinforced resin systems. Springer, Netherlands, Dordrecht
30. Saunders N (2008) Chemical reactions, 1st edn. Exploring physical science. Rosen Central/Rosen Publication, New York
31. Torgal FP, Labrincha J, Cabeza L, Goeran Granqvist C (2015) Eco-efficient materials for mitigating building cooling needs: design, properties and applications. Woodhead Publishing, Oxford

References

32. Bai J (ed) (2013) Advanced fibre-reinforced polymer (FRP) composites for structural applications, vol 46. Woodhead publishing series in civil and structural engineering. Woodhead Publishing, Oxford
33. Hollaway L (1994) Handbook of polymer composites for engineers. Woodhead Publishing Ltd, Cambridge
34. Callister WD, Rethwisch DG (2018) Materials science and engineering: an introduction, 10th edn. Wiley, Hoboken NJ
35. Silcock MD, Garschke C, Hall W, Fox BL (2007) Rapid composite tube manufacture utilizing the quicksteptm process. J Compos Mater 41(8):965–978. https://doi.org/10.1177/0021998306067261
36. Eneyew ED, Ramulu M (2014) Experimental study of surface quality and damage when drilling unidirectional cfrp composites. J Mater Res Technol 3(4):354–362. https://doi.org/10.1016/j.jmrt.2014.10.003
37. Karpat Y, Bahtiyar O, Değer B (2012) Milling force modelling of multidirectional carbon fiber reinforced polymer laminates. Procedia CIRP 1:460–465. https://doi.org/10.1016/j.procir.2012.04.082
38. Gaugel S, Sripathy P, Haeger A, Meinhard D, Bernthaler T, Lissek F, Kaufeld M, Knoblauch V, Schneider G (2016) A comparative study on tool wear and laminate damage in drilling of carbon-fiber reinforced polymers (cfrp). Compos Struct 155:173–183. https://doi.org/10.1016/j.compstruct.2016.08.004
39. Abrate S, Walton DA (1992) Machining of composite materials. part i: traditional methods. Compos Manuf 3(2):75–83
40. El-Hofy MH, Soo SL, Aspinwall DK, Sim WM, Pearson D, M'Saoubi R, Harden P (2017) Tool temperature in slotting of cfrp composites: 45th sme north american manufacturing research conference, namrc 45, la, usa. Proc Manuf 10:371–381. https://doi.org/10.1016/j.promfg.2017.07.007
41. Caggiano A (2018) Machining of fibre reinforced plastic composite materials. Materials (Basel, Switzerland) 11(3). https://doi.org/10.3390/ma11030442
42. Biron M (2014) Thermosets and composites: material selection, applications, manufacturing, and cost analysis, 2nd edn. PDL handbook series. Andrew Elsevier, Oxford
43. Lee SM (1992) Handbook of composite reinforcements. VCH, New York

Chapter 4
Advanced Methods—Vacuum Bagging and Prepreg Moulding

Abstract Vacuum bagging and prepreg moulding methods are introduced in this chapter. These advanced moulding techniques address some of the shortcomings of wet layup (hand lamination) but require additional equipment and consumables, as well as higher fabricator skill levels. The requisite equipment and consumables are described in detail. A step-by-step guide to vacuum bagging is offered and the storage and handling requirements of prepregs are considered. The advantages and limitations of these advanced methods are presented alongside their typical fibre volume (and weight) fractions. Finally, fibre volume fractions for wet layup are compared to the higher fractions typically obtained by vacuum bagging and prepreg moulding.

4.1 Introduction

The wet layup process is a simple and effective method to create FRCs. It is a versatile and low-cost fabrication method that requires minimal expertise to create a laminate.

The main problems [1–3] associated with wet layup include:

- Accuracy of fibre orientation.
- Control of the fibre and resin ratio.
- Void content (entrapped air).
- Consistency of structure or parts.
- Health and safety concerns with resins.

To overcome some of the performance issues of wet layup, vacuum bagging and prepreg moulding methods have been developed [3–7]. These advanced methods are considered here.

4.2 Vacuum Bagging

The vacuum bagging process is an advanced moulding technique that can be used alongside wet layup to remove entrapped air and provide consolidation pressure to a laminate [4, 7, 8]. A vacuum bag is made to cover the uncured laminate, air is evacuated, and atmospheric pressure is used to improve consolidation. The air evacuation process is sometimes referred to as *debulking* [2, 3].

This technique can be applied once the wet layup is complete, as a consolidation phase. Recall the six-step process in Chap. 3 (see Fig. 3.1). All steps for vacuum bagging are the same as for wet layup, except for Step 3: 'Layup and consolidation'. In step 3, consolidation is introduced via a vacuum bag. Namely, after the laminate is rolled to expel entrapped air, an additional vacuum consolidation method is applied to the wet layup.

A schematic of the vacuum bagging setup is shown in Fig. 4.1. To facilitate the application of a vacuum, extra equipment and consumables are needed. In terms of equipment, a breach unit, a vacuum pump, vacuum pot and a vacuum gauge are used. Extra consumables include a vacuum hose (tube), peel ply (optional), release film, bleeder and breather fabrics, vacuum bagging film and sealant tape (sometimes referred to as *tacky tape*). The role of the vacuum equipment and consumables is clearly outlined in Tables 4.1 and 4.2, respectively.

In this chapter, only the steps of the vacuum bagging process following completion of the wet layup stage are explained. For consistency between the techniques, the same fibre and matrix selections (described in Chap. 3) are repeated, i.e. four layers of (200 g/m^2) carbon fibre [9] and epoxy resin [10] are used. Figure 4.2a shows the fibres on the mould (flat plate), prior to the vacuum bagging assembly process. The fibres have been *wet out* using the wet layup process—the resin is still in the liquid (uncured) state.

To commence the vacuum bagging process, a release film is placed directly on top of the wet layup to prevent the consumables from adhering to the curing resin; see Fig. 4.2b. Sometimes an optional peel ply is used prior to the release film, but not in the current instance. The purpose of the peel ply is to provide a textured surface for further lamination or to facilitate secondary bonding [4, 5, 11].

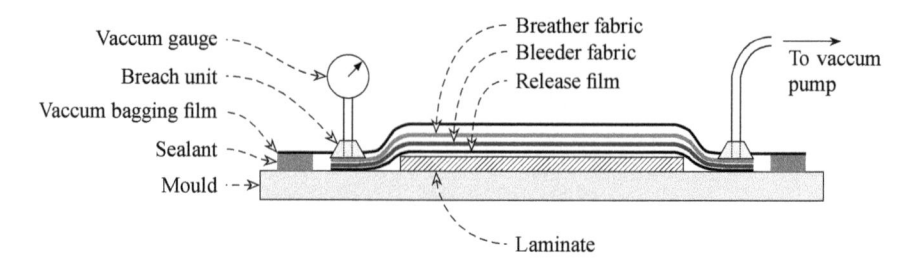

Fig. 4.1 Schematic of the vacuum bagging process

4.2 Vacuum Bagging

Table 4.1 Vacuum bagging equipment

Item	Purpose	Image
Breach unit	The breach unit allows a vacuum hose to be connected to the vacuum bagging film. Note, an alternative is to simply seal the vacuum hose between the vacuum bagging film and mould perimeter.	
Vacuum hose	Provides the link between the breach unit and the vacuum pot, as well as between the vacuum pot and vacuum pump.	
Vacuum pot (resin trap)	Prevents excess resin from entering (and hence damaging) the vacuum pump.	
Vacuum pump	A high-volume pump for drawing a vacuum.	
Vacuum (pressure) gauge	Attached to the vacuum bag via a breach unit to monitor the vacuum (consolidation) pressure.	

Table 4.2 Vacuum bagging consumables

Item	Purpose	Image
Peel ply	An optional fabric used to produce a textured surface on the cured laminate. The textured surface facilitates further lamination, secondary bonding or painting.	
Release film	A perforated sheet placed on the peel ply or directly on top of the laminate. The perforations allow the resin to 'bleed' from the laminate during the curing process. The release film prevents the other consumable from sticking to the laminate.	
Bleeder/breather fabric	Bleeder/breather fabric serves two main purposes: (i) to absorb excess resin (bleed) and reactants from the laminate; and (ii) to provide a flow path to permit the escape of air, reactants and volatiles.	
Sealant tape	Sealant tape, sometimes referred to as *tacky tape*, is used to ensure a seal between the mould and the vacuum bagging film.	
Vacuum bagging film	The vacuum bagging film encapsulates the laminate and consumables permitting a vacuum to be drawn.	

4.2 Vacuum Bagging

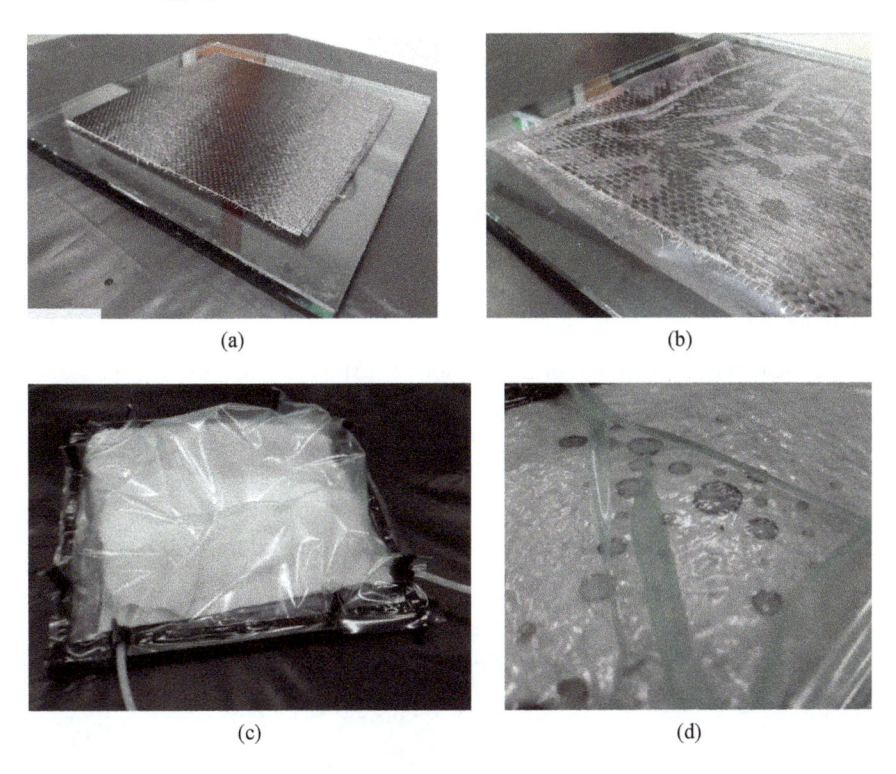

Fig. 4.2 Consolidation using a vacuum bagging process (over a wet layup): **a** final stage of the wet layup; **b** application of the release film; **c** vacuum bagging fixed with the sealant tape; and **d** extracting the excess resin under vacuum

Next, bleeder and breather fabrics are placed to cover the laminate and release film. These are often the same material [12] and are therefore (with careful control of resin absorption [3]) sometimes combined into one as illustrated in Fig. 4.2c. The combination fabric is referred to here as *bleeder/breather* fabric [5, 7, 13]. The vacuum bagging film is then sealed to the mould tool with sealant tape allowing a vacuum to be drawn within the vacuum bag. In the figure, the vacuum is drawn simply by sealing the hose between the vacuum bagging film and mould, near the perimeter. A more common method preferred by many fabricators is to use a *breach unit* (sometimes called a vacuum valve [3] or vacuum connector [5, 7]) instead of using a hose as the air outlet; see Fig. 4.1. The outcome of either methods should be identical, but the use of a hose method is preferred herein.[1]

As the vacuum is drawn, the air is evacuated and the excess resin is absorbed by the bleeder/breather fabric. In Fig. 4.2d, the excess resin can be seen as it is drawn into

---

[1] If using the vacuum hose method, to ensure a uniform consolidation pressure across the laminate and to prevent blockage, the hose can be carefully placed between two additional pieces of bleeder/breather fabric placed on top of the original fabric but to one side of the wet layup. If the hose is left on top of the layup, vacuum pressure may cause an indentation on the cured composite.

the fabric by the vacuum—without the bleeder/breather fabric, it is more difficult to evacuate the air and draw out the excess resin. A vacuum (pressure) gauge monitors the absolute pressure inside the vacuum bag and helps to identify the *leak rate*. There is no agreed threshold for an *acceptable* leak rate; the aim is simply to maintain a leak rate as close to zero as possible [14], but as a general guide less than 1 $^{mbar}/_{min}$ is recommended here as a sensible threshold target for the vacuum bagging process.

Vacuum Bagging Technique. There is more than one method to create a vacuum bag that surrounds a laminate and offers a suitable mould seal. Possible options include the following:

- The vacuum bagging film (normally nylon [2, 15]) is laid directly onto the sealant tape, which is already adhered to the mould surface [7].
- The sealant tape is stuck to the vacuum bagging film, and then the film and sealant tape are simultaneously applied to the mould tool as illustrated in Fig. 4.3.

If a mould tool is not flat but is instead either concave or convex (i.e. a female or male mould tool, respectively), additional vacuum bagging film will be needed to

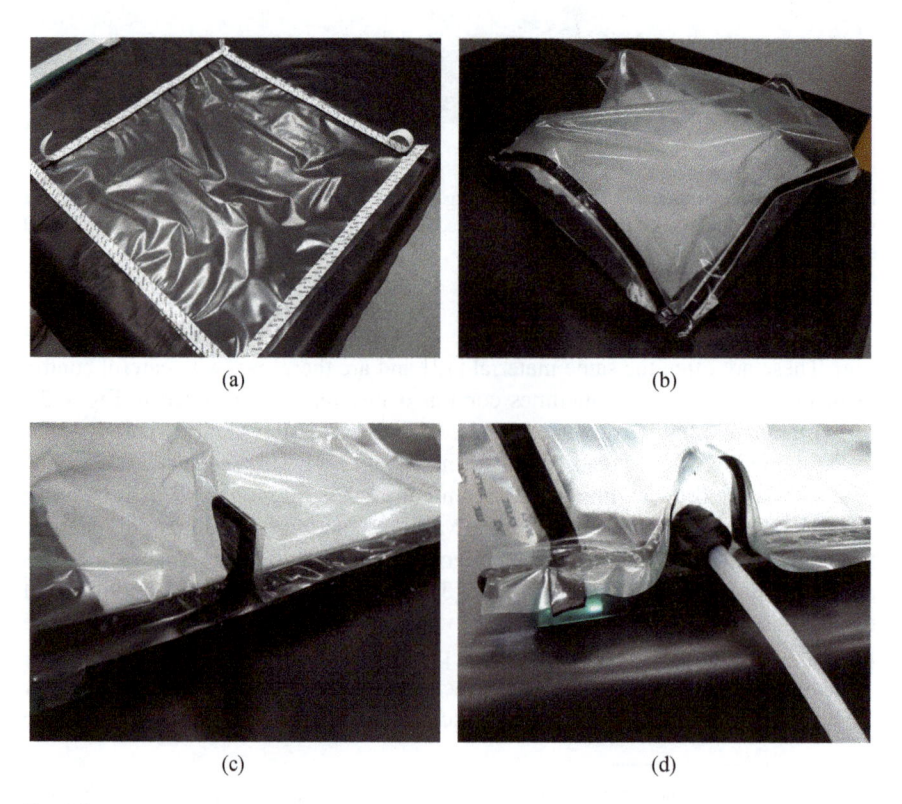

Fig. 4.3 Vacuum bagging: **a** application of the sealant tape; **b, c** and **d** steps in the vacuum bagging process

4.2 Vacuum Bagging

conform to the tool surface. As a result, there will be a need to introduce *pleats* in the bag to ensure that there is enough bagging film to conform to the contours of the mould [2]. A bag that is too large is better than one that is too small!

In Fig. 4.3, the pleated vacuum bagging process is illustrated in more detail (albeit for the flat plate mould, rather than a more complex male or female mould surface). The sealant tape is applied to the vacuum bagging film as shown in Fig. 4.3a, and then after introducing the requisite consumables, the film and sealant tape are stuck to the mould surface. The sealant tape is initially stuck to the mould tool at the four corners as shown in Fig. 4.3b, and then working from each of the corners, the pleats are formed at one or two locations on each edge; a pleat formed on one edge of the mould tool is shown in Fig. 4.3c. Pleats are normally aligned with the edges of the composite sample or geometric features. In Fig. 4.3d, the vacuum hoses are wrapped in sealant tape before being placed in their position on the perimeter of the mould as an alternative to using a breach unit (or vacuum valve/connector [3, 5, 7]). This wrapping method helps to lift the vacuum hose from the mould surface and facilitates a tight seal around the vacuum hose.

Once the vacuum bag is sealed around the mould perimeter, a vacuum is drawn and the air is extracted. The pressure inside the vacuum bag is monitored and any evident leaks are rectified; once the leaks are minimised, a leak test should be conducted for a minimum of 5 minutes [14]. The monitoring and rectification of leaks should continue until an acceptable leak rate is achieved whilst remembering that the curing process is underway. The vacuum is maintained for several hours until the composite is fully cured.

After the FRC is cured to a solid state (after 24 hours), the composite panel is demoulded. The vacuum consumables are detached, and disposed of, before the FRC panel is removed from the mould tool with a plastic wedge. The UD composite is then post-cured using the same procedure as for the wet layup panel (see Chap. 3).

Table 4.3 refines this fibre volume fraction range for the vacuum bagging process as well as offering estimates for the corresponding weight fractions (based on fibre type and form). The vacuum bagging process offers notable volume fraction improvements over the wet layup process; typical wet layup values were previously reported in Table 3.2.

The vacuum bagging process addresses some (but not all) of the issues with wet layup. The main advantages and disadvantages of vacuum bagging are outlined in Table 4.4.

Table 4.3 Typical fibre volume and weight fractions for the vacuum bagging process

Material	Random mat V_f	W_f	Woven V_f	W_f	Unidirectional V_f	W_f
Glass	0.2–0.4	0.35–0.59	0.3–0.5	0.48–0.68	0.4–0.6	0.59–0.76
Carbon	0.2–0.4	0.27–0.49	0.3–0.5	0.39–0.59	0.4–0.6	0.49–0.69
Aramid	0.2–0.4	0.23–0.44	0.3–0.5	0.34–0.55	0.4–0.6	0.44–0.64

Table 4.4 Advantages and disadvantages of vacuum bagging. Adapted from [2, 11, 16]

Advantages	Disadvantages
High fibre volume fractions. Air and excess resin are drawn from the laminate resulting in higher fibre volume fractions and lower void content than for wet layup.	*Fibre orientation*. The fibre orientation is still manually aligned by the fabricator, and hence accuracy and consistency is still a potential issue.
Improved consolidation. Laminate consolidation and wet out are improved due to pressure and resin flow, and hence a more uniform thickness and a better surface finish can be attained.	*Fabricator skill levels*. Fabricators require higher skill levels and there can be difficulty in bagging complex shapes.
Health and safety. The vacuum bag minimises exposure to volatile organic compounds during curing, but the resin mixing process is still a manual task.	*Increased process costs*. There are extra equipment and consumable costs, and vacuum tight moulds are needed.

4.3 Prepreg Moulding

Prepregs are pre-impregnated materials usually comprising carbon fibre and epoxy. They are supplied on a plastic backed roll and are stored in a protective bag in a freezer (at −18 °C) [3]. The prepregs are in a stable intermediate cure stage (partially crosslinked) that only requires heat to cure. Prepregs can be stored in the freezer for 3–12 months [15], but at room temperature the prepreg can remain uncured for a shorter period of time (1–2 months is typical) [17]. The time a prepreg can be stored in the freezer is referred to as the *freezer-life* [15, 18] or *shelf-life* [14], whilst the term *out-time* [2, 3] or *out-life* [18] identifies the time a prepreg can remain at room temperature before it is no longer useable.

Prior to heating, the prepregs are removed from the freezer and protective bag, cut to size and left to thaw at room temperature for several hours to prevent condensation issues [2]. The prepreg roll is resealed and returned to freezer storage as soon as possible to maintain the shelf-life [3]. Cutting can be performed with a sharp knife or scissors. Once the plastic backing is removed, the prepreg is ready for layup. The freezer and out-life of prepregs need to be carefully monitored to ensure appropriate storage and handling times are not exceeded.

At room temperature, prepregs are *tacky* and are easily handled with safety gloves [7] as shown in Fig. 4.4. To help conformation to the mould and to increase tack and drape during layup, prepregs can be gently heated [4, 5]. To optimise performance, the fibre and resin ratios are predefined by the supplier with near-net fibre volume fractions [2]. Moreover, they are easily cut in the uncured state [7] and hence, precut patterns can simplify prepreg layup. The tackiness allows for easy positioning in the mould and hence provides maximum accuracy in fibre orientation. The material costs for prepregs, however, are significantly higher (perhaps 4 times higher [19]) than those for wet layup and small quantities cannot often be purchased from supplies

4.3 Prepreg Moulding

Fig. 4.4 Prepreg: woven carbon fibres in a partially cured epoxy

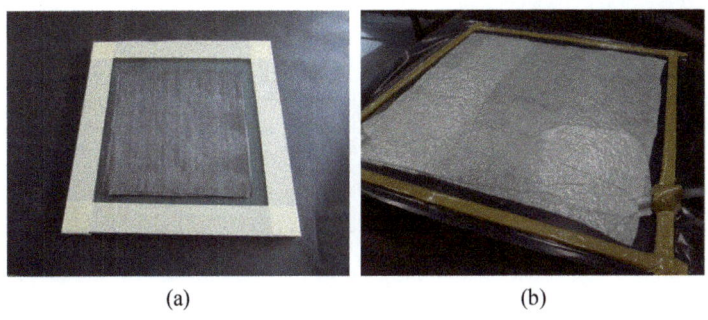

Fig. 4.5 Prepreg moulding process: **a** prepreg layup (4 plies); and **b** vacuum bagging and debulking

who tend to prefer not to 'split' a roll. They are therefore often only readily used in the highest performance industries (e.g. aerospace and sports equipment sectors [3]).

The prepreg layup and consolidation is shown in Fig. 4.5. During layup, the prepreg [20] is laid on a pre-released mould, rolled after each ply and then consolidated in a vacuum bag (see method in Sect. 4.2). Prepregs cure at elevated temperatures, and therefore the mould release and the vacuum bagging consumables (release film, bleeder/breather fabric, vacuum bagging film and sealant tape) all need to be suitable at the cure temperature. There are three main temperature ranges for prepregs: low (60 °C); medium (120 °C); and high (180 °C). However, some prepregs (including the one used here) allow a variable cure temperature with modified curing times—lower curing temperatures require longer curing times than higher temperatures.

To cure prepregs, the temperature is typically ramped up from room temperature, held at the cure temperature in an oven for the requisite time (sometimes referred to as the *dwell time*) and then cooled down. The temperature ramp-up and cool-down rates are provided by the prepreg supplier, but are normally only a few degrees per

Table 4.5 Typical fibre volume and weight fractions for prepreg processes

Material	Random mat		Woven		Unidirectional	
	V_f	W_f	V_f	W_f	V_f	W_f
Glass	N/A	N/A	0.4–0.55	0.58–0.72	0.5–0.7	0.68–0.83
Carbon	N/A	N/A	0.4–0.55	0.49–0.64	0.5–0.7	0.59–0.77
Aramid	N/A	N/A	0.4–0.55	0.44–0.59	0.5–0.7	0.55–0.74

Table 4.6 Advantages and disadvantages of prepreg moulding. Adapted from [2, 16]

Advantages	Disadvantages
High fibre volume fractions. Fibre and resin ratios are predefined ensuring high fibre volume fraction and low void content.	*Fibre orientation.* The fibre orientation is still aligned by the fabricator but there is potential for automation.
Longer working times. The extended working times mean complex layups can be achieved without concerns about resin working (curing) times.	*Fabricator skill levels.* Fabricators require higher skill levels—similar to vacuum bagging.
Shorter curing times. Prepreg cure times can be much shorter than for wet layup or vacuum bagging—see Table 4.5.	*Moulding costs.* Prepreg costs are high and storage controls are needed to monitor freezer and out-life. Elevated cure temperatures mean tooling needs are increased.
Health and safety. Health and safety issues are minimised using prepregs.	

minute (for example, 2 – 3 °C/min). The laminate is maintained under vacuum pressure during the entire curing process until it is removed from the oven, demoulded and post-processed. Here, a dwell time of 5 hours at a temperature of 80 °C is employed.

Table 4.5 provides fibre volume and weight fraction estimates (based on fibre type and form) for prepreg moulding. The premium cost (and performance) associated with prepreg means that random mat carbon fibre prepregs are not readily available— it is not sensible to pay a premium cost for a suboptimal solution. Volume fractions for prepreg moulding are towards the highest levels that can be sensibly realised in practice.

The main advantages and disadvantages of prepreg moulding are outlined in Table 4.6.

4.4 Summary

Advanced moulding methods require additional equipment and consumables but address many of the shortcomings of wet layup. They offer higher fibre volume fraction (and hence weight fraction) laminates with lower void content. The result is therefore a stiffer and stronger laminate, but moulding costs for advanced methods are

4.4 Summary

Table 4.7 Typical fibre volume fractions for wet layup and advanced moulding methods

Method	Fibre form		
	Random (V_f)	Woven (V_f)	Unidirectional (V_f)
Wet layup*	0.1–0.3	0.2–0.4	0.3–0.5
Vacuum bagging	0.2–0.4	0.3–0.5	0.4–0.6
Prepreg moulding	N/A	0.4–0.55	0.5–0.7

*Wet layup values are restated from Chap. 3 for convenience

higher than for wet layup and necessitate higher fabricator skill levels. Prepreg moulding has the highest cost for the methods described. Typical fibre volume fractions for vacuum bagging and prepreg moulding are compared to wet layup in Table 4.7.

4.5 Questions

Vacuum Bagging

Question 4.1 The main problems associated with wet layup were listed as:

- Accuracy of fibre orientation.
- Control of the fibre and resin ratio.
- Void content (entrapped air).
- Health and safety concerns with resins.

Which of these are *addressed*, *partially addressed* or *not addressed* by

a Vacuum bagging.
b Prepreg moulding.

Question 4.2 Describe the role of each item of equipment in the vacuum bagging process:

a Breach unit.
b Vacuum hose.
c Vacuum pot.
d Vacuum pump.
e Vacuum gauge.

Question 4.3 Identify the five consumables in Fig. 4.6. Describe their role in the vacuum bagging process.

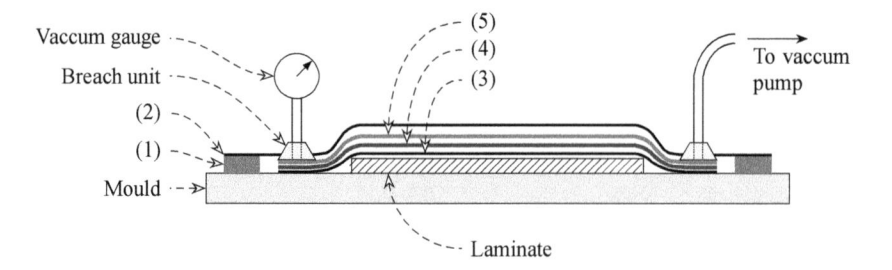

Fig. 4.6 Consolidation using a vacuum bagging process (over a wet layup)

Question 4.4 Provide fibre volume fraction and fibre weight fraction estimates (i.e. the range of values) for each fibre reinforcement using the vacuum bagging process:

a Unidirectional carbon fibre.
b Plain weave carbon fibre.
c Twill weave glass fibres.

Question 4.5 A vacuum bag is used to remove air trapped between the fibres and to consolidate the laminate. What term is used to describe the process of air evacuation?

Prepreg Moulding

Question 4.6 The term *prepreg* is an abbreviation for what term?

Question 4.7 Provide fibre volume fraction and fibre weight fraction estimates (i.e. the range of values) for a unidirectional carbon fibre prepreg.

Question 4.8 The room temperature is 24 °C, and the ramp-up and cool-down rates are scheduled to be 2 °C per minute. What will the total manufacturing time be, if the prepreg is cured for 5 hours at 80 °C?

4.6 Problems

Problem 4.1 A unidirectional carbon fibre-reinforced epoxy composite was produced by wet layup in Problem 3.4. If the same constituents (fibres and resin) are used to fabricate a composite using the vacuum bagging process, estimate the strength of the FRC loaded in axial tension. Compare the tensile strength of the wet layup and vacuum bagged laminates—is there a significant difference? If so, why? Recall, the tensile strength of the carbon fibres and the epoxy matrix were assumed to be 3800 MPa and 70 MPa, respectively.
Answer. $\sigma_{cl}^* = 1520$ MPa (wet layup), $\sigma_{cl}^* = 1900$ MPa (vacuum bagging). Why? V_f.

4.6 Problems

Problem 4.2 Which of the following materials and manufacturing processes would you expect to be suitable to achieve a tensile modulus of 80 GPa and a tensile strength of 1350 MPa?

- Wet layup of unidirectional HS carbon fibres (40 vol%) in an epoxy resin.
- Vacuum bagging of unidirectional HM aramid fibres (55 vol%) in an epoxy resin.
- Prepreg moulding of unidirectional (an E-glass fibres (60 vol%) in an epoxy matrix.

The elastic modulus and strengths for E-glass, HS carbon and HM aramid fibres, and epoxy are given in Table 4.8.
Answer. Wet layup ✓, Vacuum bagging ✗, prepreg moulding ✗.

Table 4.8 Fibre and resin properties

Material	Density (g/cm^3)	Elastic modulus (GPa)	Tensile strength (MPa)
Glass (E-glass)	2.50–2.62	69–81	2000–3450
Carbon (HS)	1.75–1.80	228–300	3400–7100
Aramid (HM)	1.40–1.44	130–131	3600–4100
Epoxy	1.1–1.4	2.41–4.1	27.6–130

References

1. Karlsson KF, TomasÅström B (1997) Manufacturing and applications of structural sandwich components. Compos Part A: Appl Sci Manuf 28(2):97–111. https://doi.org/10.1016/S1359-835X(96)00098-X
2. Lee SM (1992) Handbook of composite reinforcements. VCH, New York
3. Strong AB (2008) Fundamentals of composites manufacturing: materials, methods and applications, 2nd edn. Society of Manufacturing Engineers, Dearborn, Mich
4. Astrom BT (2018) Manufacturing of polymer composites, 2nd edn. Routledge, Boca Raton
5. Barbero EJ (2017) Introduction to composite materials design, 3rd edn. Composite materials. CRC Press, Boca Raton
6. Mayer RM (1993) Design with reinforced plastics: a guide for engineers and designers. Springer, Netherlands, Dordrecht. https://doi.org/10.1007/978-94-011-2210-8
7. Wanberg J (2010) Composite materials: fabrication handbook #2. Composite garage series. Wolfgang Publications, Stillwater, Minnesota
8. Kutz M (2002) Handbook of materials selection. Wiley, New York and Chichester
9. Hyosung Corporation (2017) Tansome carbon fiber. https://www.hyosungusa.com/files/advanced/tansome_catalog_2017.pdf
10. Fiber Glass International (2021) Es180 epoxy data system. https://martglass.com.au/wp-content/uploads/FGI-Epoxy-Resin.pdf
11. Campbell FC (2010) Structural composite materials. ASM International, Materials Park, Ohio
12. Sheikh-Ahmad JY (2009) Machining of polymer matrix composites. Springer, New York and London

13. Lubin G (ed) (2013) Handbook of composites. Springer, New York, NY
14. Summerscales J, Graham-Jones J, Pemberton R (2018) Marine composites: design and performance. Woodhead Publishing series in composites science and engineering. Woodhead Publishing, Oxford
15. Campbell FC (2004) Manfacturing processes for advanced composites. Elsevier Advanced Technology, Oxford
16. Gurit (2019) Guide to composites. https://www.gurit.com/Our-Business/Composite-Materials
17. Brain NG (1991) Composites: tooling and component processing. iSmithers Rapra Publishing
18. Baker AA, Dutton S, Kelly D (2004) Composite materials for aircraft structures, 2nd edn. AIAA education series. American Institute of Aeronautics and Astronautics, Reston, VA
19. Adams PJ (2000) Revolutionary materials: technology and economics: 32nd International SAMPE Technical Conference, Boston Park Plaza Hotel, Boston, Massachusetts, November 5–9 2000, International SAMPE Technical Conference series, vol 32. Society for the Advancement of Material and Process Engineering, Covina, Calif
20. Solvay (2021) Technical data sheet: Vtm® 260 series (prepeg). https://catalogservice.solvay.com/

Chapter 5
Composite Testing—How Accurate Are Design Estimates?

Abstract A selection of simple composite design (mechanics) equations were introduced in Chap. 2 to enable prediction of the mechanical properties of fibre-reinforced composites (FRCs). The rule of mixtures (RoM) and inverse rule of mixtures (IRoM) expressions were introduced for elastic moduli estimation, whilst longitudinal strength was considered via the Kelly-Tyson (KT) model. Chapters 3 and 4 then introduced wet layup, and vacuum bagging and prepreg moulding methods for composite manufacture. A carbon fibre-reinforced panels was fabricated using *similar* constituents (i.e. 200 g/m² carbon fibres in an epoxy matrix) for each fabrication method. This chapter builds on these earlier chapters via a comparison of design (mechanical property) estimates and experimental data. Tensile tests are performed on multiple specimens cut from each of the panels. This chapter provides guidance on the accuracy of design estimates and offers support for the typical (rule of thumb) volume fractions suggested in Chaps. 3 and 4 for each manufacturing approach.

5.1 Introduction

In previous chapters, unidirectional (UD) carbon fibre-reinforced epoxy composites were fabricated using basic wet layup (Chap. 3), and vacuum bagging and prepreg moulding (Sects. 4.2 and 4.3). The same six steps were followed to create *similar* carbon fibre panels. The laminates were made from similar constituents, i.e. the fibre type and form, and the matrices were all similar. Nonetheless, the laminate performance will differ in each case as a result of the variations in material properties and manufacturing processes.

The fibre volume fraction (V_f) is of critical importance to the mechanical performance of FRCs [1] and is directly affected by the fabrication method. An understanding of V_f (for each method) is therefore an important consideration for a composite designer. Here, measured fibre volume fractions are used to inform design estimates that are compared to tensile experiments for specimens cut from each of the carbon fibre panels. In doing so, the chapter offers insight into the accuracy of the composite design equations (see Chap. 2) and provides validation for the rule of thumb values used to estimate fibre volume fractions for each of the manufacturing methods—see Table 4.7.

Design estimates are a fundamental step in a product development process as they enable the designer to make structural predictions (within margins of error) prior to prototype testing, but it should be noted that structural predictions are not a substitution for physical testing. After design calculations are complete, a prototype should always be fabricated and tested to the relevant product or engineering standards.

5.2 The Tensile Test

The uniaxial tensile test is a common method used to determine the characteristic mechanical properties of materials [2], including composites [3].[1] Tensile test data from token specimens is frequently used to provide elastic moduli, tensile strength and fracture strain measurements that are subsequently used to inform structural design decisions [5].

To perform a tensile test, the specimen is held in a uniaxial test machine and monotonically loaded (see Fig. 5.1) until failure occurs [4]. During a tensile test, force (load) and extension (deformation) are recorded. A load cell on the machine records the force, whilst the deformation of a specimen is typically measured using a clip-on extensometer (or sometimes bonded strain gauges) rather than via the machine's own *crosshead* displacement measurements; crosshead displacements are

Fig. 5.1 Tensile test of a carbon fibre-reinforced composite

---

[1] Tensile tests are usually performed in accordance with a recognised testing standard; for FRCs, ASTM D3039 [4] is often used.

5.2 The Tensile Test

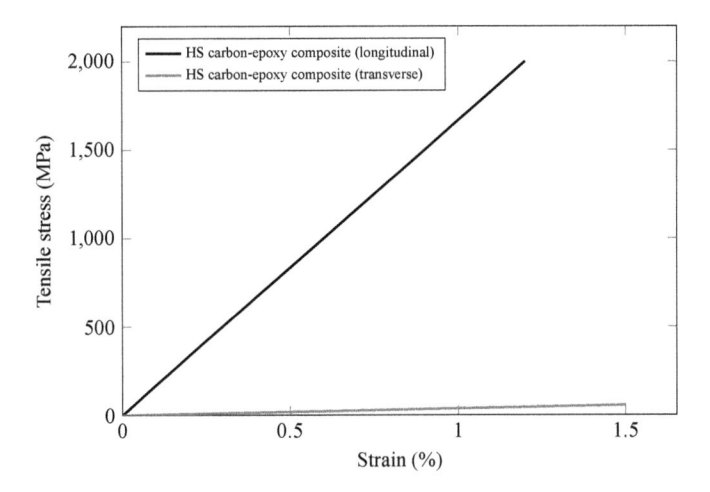

Fig. 5.2 Tensile stress-strain curve for a UD carbon fibre composite in longitudinal and transverse directions

not usually used as the measurements need to be calibrated based on the machine compliance [6].

The load and deformation data is converted to engineering stress (σ) and engineering strain (ε), via [2]

$$\sigma = \frac{F}{A} \tag{5.1}$$

and

$$\varepsilon = \frac{\Delta L}{L} \tag{5.2}$$

where F is the force, A is the cross-sectional area of the specimen, ΔL is the change in length (extension) and L is the original length of the specimen or the extensometer gauge length in the tensile test.

A typical tensile stress-strain curve for a unidirectional carbon FRC with fibres aligned parallel (longitudinal) and transverse (perpendicular) to the loading direction is shown in Fig. 5.2.

The tensile modulus (stiffness) of the material sample is determined via a correlation of stress and strain. To calculate the stiffness, a *submaximal* load is correlated to the deformation. As can be seen in Fig. 5.2, the tensile stress is proportional to strain and hence, the elastic modulus of the sample is the constant of proportionality, viz. [2]

$$E = \frac{\sigma}{\varepsilon} \tag{5.3}$$

where E_{cl} could be used in place of E to denote axial (longitudinal) modulus for a composite sample, and E_{ct} could be used for the transverse modulus.

The tensile strength of the FRC (σ^*) is the stress at failure (i.e. the maximum stress) given by

$$\sigma^* = \frac{F_{max}}{A} \tag{5.4}$$

in which σ_{cl}^* could replace σ^* for longitudinal strength, and σ_{ct}^* could be interchanged for transverse strength.

5.3 Fibre-Reinforced Composite Specimens

The composite panels fabricated in Chaps. 3 and 4 were used to create the tensile test specimens; longitudinal and transverse samples were created for each fabrication method. Here, more longitudinal specimens were created from the plate than transverse specimens as the mechanical properties of the former are usually of greater significance to the designer—as previously mentioned, the transverse stiffness and strength of unidirectional laminates are much lower than those in the longitudinal direction.

The tensile test specimens were cut from the panel into rectangular strips approximately 20 mm wide. At least five longitudinal and three transverse specimens were cut from each panel.[2] An example specimen is shown for each fabrication method in Fig. 5.3.

The thickness of the FRC specimens is dependent upon the manufacturing method. The wet layup specimens are the thickest, whilst the vacuum bagging and prepreg specimens are thinner. A thicker specimen has a lower fibre volume fraction (and hence, a higher matrix fraction) since the same volume of carbon fibres are used in each fabrication method. The lowest fibre volume fractions are therefore evident in the wet layup specimens, whilst the vacuum bagging and prepreg specimens have higher volume fractions. A laminate's thickness is related to its fibre volume fraction [7] via

$$t = \frac{n A_w}{\rho_f V_f} \tag{5.5}$$

where t is the test specimen thickness, n is the number of layers, A_w is the areal weight of the fabric, ρ_f is the mass density of the fibre used and V_f is the fibre volume fraction.

The mean dimensions of the tensile test specimens and the fibre volume fractions are given in Table 5.1. In the table, V_f from Eq. (5.5) is calculated based on the average of five thickness measurements (measured prior to testing) from the successful test specimens (see Sect. 5.4) using $\rho_f = 1800 \, \text{kg/m}^3$ [8, 9]. The *typical* (rule of thumb) fibre volume fractions previously reported (for unidirectional laminates in Table 4.7)

---

[2] It is common practice to report mean measurements for at least five successful test specimen.

5.3 Fibre-Reinforced Composite Specimens

Fig. 5.3 Tensile specimens—wet layup, vacuum bagging and prepreg moulding (left to right)

Table 5.1 Tensile test specimens: thickness and fibre volume fraction

Manufacturing method	Specimen thickness (mm)	V_f from Eq. (5.5)	V_f typical (rule of thumb)
Wet layup	1.32	0.34[a]	0.3–0.5
Vacuum bagging	0.89	0.50	0.4–0.6
Prepreg moulding	0.83	0.54	0.5–0.7

[a]Wet layup estimates in Example 3.1 assumed a slightly higher V_f ($= 0.4$) than the measured value of 0.34

are also included in the table for comparison. Note the correlation between the typical fibre volume fractions for each fabrication method and the corresponding calculated value.

5.4 Tensile Properties: Elastic Moduli and Strength

Testing Method. Tensile tests data was produced with an Instron Universal Testing System (UTS) using displacement control. The strain rate was maintained at 2 mm/min (monotonic loading) until failure, in accordance with ASTM D3039 [4]. A clip-gauge extensometer (with a 50 mm gauge length) was used to accurately measure gauge length deformation. The stresses and strains observed during testing are derived from the force-deformation data and the specimen dimensions using Eqs. (5.1) and (5.2), respectively.

Results and Discussion. A typical plot of the stress-strain measurements for each manufacturing method (wet layup, vacuum bagging and prepreg moulding) is shown

Fig. 5.4 Typical stress-strain curves: **a** longitudinal; and **b** transverse loading

(a)

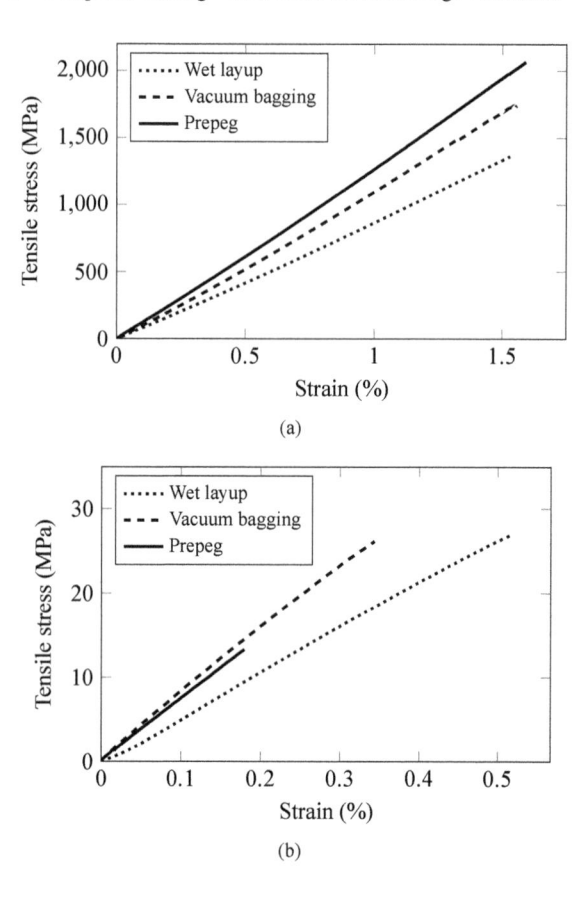

(b)

in Fig. 5.4. The measurements for the longitudinal and transverse fibre orientations for each fabrication method are plotted on the same graph to clearly illustrate the significant performance difference.

The mean elastic moduli and tensile strength measurements are compared to design estimates (based on the RoM, IRoM and KT models) in Table 5.2. The design estimates are calculated using carbon fibre properties taken from supplier data sheets—for the fibres used in the wet layup and vacuum bagging processes, E_f and σ_f^* are 250 GPa and 5516 MPa [8], whilst E_f and σ_f^* for the prepreg fibres are 230 GPa and 4900 MPa [10]. E_m and σ_m^* are assumed (in all estimates) to be 3.0 GPa and 50 MPa, respectively. The fibre volume fractions V_f are taken from Table 5.1. As a further comparison, Table 5.2 shows supplier data for the prepreg material, normalised to 60% V_f [9].

In Table 5.2, the number of specimens successfully tested are identified in round brackets. A successful test is identified by a specimen fracture in the gauge length region. Thinner specimens tend to be more prone to premature fracture in the jaws (even if *tabs* are used) and hence fewer prepreg samples were successfully recorded. Transverse properties tend to be more difficult to accurately measure than longitudi-

5.4 Tensile Properties: Elastic Moduli and Strength

Table 5.2 Tensile test specimens: estimated and measured mechanical properties

Manufacturing method	Orientation	Elastic modulus (mean, GPa) Experimental	Estimated	Prepreg 60%V_f	Tensile strength (mean, MPa) Experimental	Estimated	Prepreg 60%V_f
Wet layup	Longitudinal	80.0 (5)*	87.0 [8.8]†	–	1310.4 (5)	1875.4.0‡ [43.1]	–
	Transverse	5.7 (3)	4.5 [−21.1]	–	40.6 (3)	≪ 50.0	–
Vacuum bagging	Longitudinal	117.4 (5)	126.5 [7.8]	–	1855.0 (5)	2758.0 [48.7]	–
	Transverse	7.5 (3)	5.9 [−21.3]	–	25.3 (3)	≪ 50.0	–
Prepreg moulding	Longitudinal	126.2 (3)	125.6 [−0.5]	131.0	1967.5 (3)	2499.0 [27.0]	2575.0
	Transverse	7.2 (1)	6.4 [−11.1]	9.1	13.6 (1)	≪ 50.0	40.0

[a] Number of specimens are printed in round brackets.
† Percentage relative errors are printed in square brackets.
‡ KT strength is calculated using fibre strength *only*, assuming $\sigma_f^* = 5516$ MPa (wet layup and vacuum bagging) or 4900 MPa (prepreg moulding).

nal specimens—particularly transverse strengths. During clamping, stresses can be induced that cause the more fragile specimens to fracture.

In considering the results, we can conclude... Design estimates (based on RoM, IRoM and KT models) offer the first approximation of the mechanical properties of FRCs. In practice, design and modelling assumptions along with manufacturing imperfections mean that estimates are likely (at best) to offer only an appraisal of the *real* situation, rather than a perfect agreement.

Elastic Moduli. In Table 5.2, the RoM and IRoM estimates provide a reasonably accurate prediction of the elastic moduli—as confirmed by others [1, 11]. In terms of the longitudinal elastic moduli, the experimental measurements are in excess of 90% of the design estimates. Transverse moduli estimates are less accurately predicted than their longitudinal counterparts but still lie within a reasonable margin of error. Moreover, the accuracy of transverse estimates can (as previously mentioned in Sect. 2.3) be improved using the more complex, semi-empirical Halpin-Tsai expression in Ref. [12].

Tensile Strength. The longitudinal and transverse strengths of FRCs tend to be more difficult to accurately predict than those for the elastic moduli since the failure mechanisms are influenced by many factors [1]. The KT estimates in Table 5.2 are calculated using mean fibre bundle strengths from supplier data [8, 10]. This method is simple but tends to offer a significant overestimate of composite performance as it does not consider the nonuniformity of fibre strengths. In reality, fibre strengths are dependent upon the presence of flaws along their length [1]. These flaws cause a considerable spread of fibre strengths about a mean value [11, 13]. Moreover, fibre damage can occur during handling [11] and fabric processing [14]. In tension, the weaker fibres will fail first resulting in a local stress concentration in the adjacent unbroken fibres, which in turn increases the likelihood of their failure [15]. The use of a mean fibre bundle strength is therefore likely to offer an upper bound strength estimate whilst a lower bound estimate can be taken as the stress at the first fibre failure. To improve the longitudinal strength estimate therefore requires consideration of the strength variability. Here, an arbitrary fibre strength factor of 0.7 in the KT model yields estimates that closely correlate to our experiments: wet layup = 1312.8 MPa (0.2%); vacuum bagging = 1930.6 MPa (4.1%); and prepreg = 1852.2 MPa (−5.9%).

In this chapter, tensile properties (only) have been reviewed. As an exercise for the reader, it would be worthwhile to benchmark the design estimates suggested for compression and shear performance (see Chap. 2). The measurement of compressive properties for unidirectional composites (see ASTM D3410 [16] and D6641 [17]) is more difficult than for tensile properties [18], but a simple flexural test (ASTM D7264 [19]) can be used to approximate the compressive strength (values can differ by ≈10–20%) [20]. The in-plane shear properties (shear modulus and strength) can be measured using a uniaxial tensile test with a ±45° laminate—see ASTM D3518 [21].

5.5 Factor of Safety (FoS)

In a practical sense, design estimates from the RoM, IRoM and KT models offer the designer a starting point for structural predictions. However, to ensure composite failure does not occur in service and to offer a safety net, it is usual to make a structure stronger than needed. This approach is justified by the fact that there is some level of uncertainty in the material properties (e.g. fibre properties), design calculations, manufacturing processes and the mechanical loads that a structure will experience in service. To account for the level of uncertainty in the calculations, a FoS is often specified.

The FoS is defined as the maximum allowable stress (or failure strength) of the composite structure (σ_c^*) divided by the composite's design (or working) stress (σ_c), viz.

$$\text{FoS} = \frac{\sigma_c^*}{\sigma_c} \tag{5.6}$$

The FoS is selected as a positive real number. The chosen value depends on several factors, but two of the most important are:

- The accuracy to which structural loads and material properties can be estimated or characterised.
- Consequences that would result from failure of the composite structure or part (severity of the application).

An excessively large FoS results in an overdesign, and hence, the chosen values are usually between 1.2 and 4.0 [2]. A higher FoS is usually applied to FRCs than to traditional (isotropic) materials and therefore values ≥ 2.0 tend to be considered the norm.

5.6 Summary

The RoM, IRoM and KT models offer the first approximations of the mechanical properties (elastic moduli and strength values) of FRCs.

Design estimates for elastic moduli are more accurately predicted than composite strengths. Fibre strengths are dependent upon the presence of flaws and there can be considerable variation in the mean value. Moreover, fibre processing and composite fabrication methods can cause damage to pristine fibres.

The thickness of a FRC is related to fibre volume fraction (V_f) via

$$t = \frac{nA_w}{\rho_f V_f}$$

In practice, model assumptions and manufacturing imperfections mean design estimates are unlikely to offer perfect agreement with experimental measurements.

Thus, the designer is responsible for the selection of an appropriate safety net (i.e. factor of safety, FoS) to compensate for any potential errors, as well as for design uncertainties and other in-service practical implications outside of the idealised test cases used.

The FoS is defined as

$$\text{FoS} = \frac{\sigma_c^*}{\sigma_c}$$

Factor of safety values are chosen based on the application and usually range from 1.2 to 4.0.

5.7 Questions

Questions 5.1 In a tensile test, a specimen is monotonically loaded until failure. What measurements are usually recorded?

Questions 5.2 What is an *extensometer* and why is it used? What could be used instead of an extensometer?

Questions 5.3 How is the tensile modulus determined from the measured data? What about the tensile strength?

Questions 5.4 Design (RoM, IRoM and KT) estimates tend to offer an approximation of the tensile properties of FRCs. The estimations are not exact. Why?

Questions 5.5 What is a *factor of safety* (FoS)? Why is it important when designing composite structures?

5.8 Problems

Problem 5.1 A tensile test is conducted on a unidirectional glass-polyester composite. The stress-strain curve is shown in Fig. 5.5.

a. What is the tensile modulus of the specimen?
b. What are the longitudinal strength and failure strain?

Answer. $E = 35\,\text{GPa}$, $\sigma_{cl}^* = 1000\,\text{MPa}$, $\epsilon_{cl}^* = 2.85\%$.

Problem 5.2 If the composite in Problem 5.1 is fabricated using 6 plies of 300 g/m² glass fibres and the specimen is 1.41 mm thick, what is the fibre volume fraction? Assume $\rho_g = 2550\,\text{kg/m}^3$.
Answer. $V_f = 0.5$.

5.8 Problems

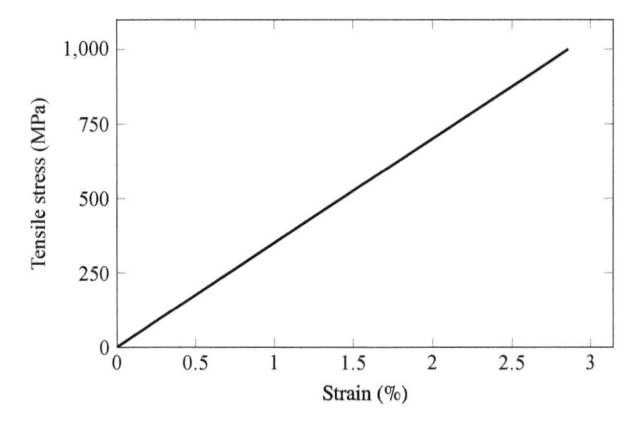

Fig. 5.5 Tensile stress-stain curve of the glass-polyester composite

Problem 5.3 A unidirectional carbon fibre-reinforced composite has a tensile strength of 1800 MPa (experimentally measured). Assuming the member is 20 mm wide:

a. What is the minimum thickness that is required to support a tensile load of 35 kN?
b. What thickness would be required if a FoS of 2.0 is selected for the application?

Answer. a. 0.97 mm, b. 1.94 mm.

Problem 5.4 Sketch the stress-strain curves for a HS carbon fibre-reinforced epoxy composite with $V_f = 0.45$. Assume the carbon fibre and epoxy have elastic moduli of 245 GPa and 3.4 GPa, and tensile strengths of 4900 MPa and 60 MPa, respectively.

References

1. Hull D, Clyne TW (1996) An introduction to composite materials, 2nd edn. Cambridge solid state science series. Cambridge University Press, Cambridge
2. Callister WD, Rethwisch DG (2018) Materials science and engineering: an introduction, 10th edn. Wiley, Hoboken
3. Strong AB (2008) Fundamentals of composites manufacturing: materials, methods and applications, 2nd edn. Society of manufacturing engineers, Dearborn, Mich
4. ASTM (2017) Standard test method for tensile properties of polymer matrix composite materials
5. Smith WF, Hashemi J, Presuel-Moreno F (2019) Foundations of materials science and engineering, 6th edn. McGraw-Hill Education, New York
6. Huerta E, Corona JE, Oliva AI, Avilés F, González-Hernández J (2010) Universal testing machine for mechanical properties of thin materials. Revista mexicana de física 56(4):317–322
7. Curtis PT (1988) Crag test methods for the measurement of the engineering properties of fibre reinforced plastics

8. Hyosung Corporation (2017) Tansome carbon fiber. https://www.hyosungusa.com/files/advanced/tansome_catalog_2017.pdf
9. Solvay (2021) Technical data sheet: Vtm(mohana)series (prepeg). https://catalogservice.solvay.com/
10. Toray Inc (2021) Technical data sheet: T700s no. CFA-005. http://stg.toray.testcrafting.com
11. Agarwal BD, Broutman LJ, Chandrashekhara K (2006) Analysis and performance of fiber composites, 3rd edn. Wiley and Chichester, Hoboken
12. Halpin JC, Tsai SW (1967) Environmental factors in composite design. Air force materials laboratory
13. Virk AS, Summerscales J, Hall W, Grove SM, Miles ME (2009b) Design, manufacture, mechanical testing and numerical modelling of an asymmetric composite crossbow limb. Compos B Eng 40(3):249–257. https://doi.org/10.1016/j.compositesb.2008.10.004
14. Lee B, Leong KH, Herszberg I (2001) Effect of weaving on the tensile properties of carbon fibre tows and woven composites. J Reinf Plast Compos 20(8):652–670. https://doi.org/10.1177/073168401772679011
15. Zweben C (1968) Tensile failure analysis of composites. AIAA J 2:2325
16. ASTM (2008) Standard test method for compressive properties of polymer matrix composite materials with unsupported gage section by shear loading
17. ASTM (2009) Standard test method for compressive properties of polymer matrix composite materials using a combined loading compression (CLC) test fixture
18. Adams D (2019) Optimum unidirectional compression testing of composites. www.compositesworld.com/articles/optimum-unidirectional-compression-testing-of-composites
19. ASTM (2015) Standard test method for flexural properties of polymer matrix composite materials
20. Adams D (2017) Can flexure testing provide estimates of composite strength properties? www.compositesworld.com/articles/can-flexure-testing-provide-estimates-of-composite-strength-properties
21. ASTM (2007) Standard test method for in-plane shear response of polymer matrix composite materials by tensile test of a ±45° laminate

Chapter 6
Moulding Composite Parts

Abstract So far, this book has focused on flat, planar laminates in order to simply and easily illustrate composite design and manufacture considerations. However, creating most composite structures (or parts) requires a mould. This chapter considers composite design in the context of part *mouldability* (i.e. design for manufacture, DFM), and offers the composite designer some guidance on mould design and construction. In terms of part mouldability, special attention is given to draft angles (and undercuts), surface textures and sharp corners. Mould design and construction are examined in the context of mould durability; speed and ease of construction; and the mould tool fabrication cost. A *step-by-step* mould-making process is presented to enable the reader to better understand moulding considerations for laminates. A simple composite part (a frisbee) is then moulded to illustrate the fabrication process.

6.1 Introduction

To make composite panels with a smooth surface on one side (on the mould side), a flat plate was previously used as a mould surface. The same flat plate was used to make composites using wet layup (Chap. 3), and vacuum bagging and prepreg moulding methods (Chap. 4) [1–3]. Of course, not all composite laminates flat surfaces. To create curved composite structures, or to reproduce a composite part numerous times, a mould tool is usually needed [4]. A female mould can be used to create a laminate with a smooth convex surface on the composite structure have, whilst a male mould can be used to create a laminate with a smooth concave surface. Matched moulds can be used to create a smooth surface on both the inner (concave) and outer (convex) composite surfaces, i.e. a laminate with two finished surfaces [5, 6].

In this chapter, we focus on composite part design and how to make mouldable fibre-reinforced composite (FRC) parts, including mould design and construction—for further information on mould design, the reader is directed to references [5, 7], whilst mould- making examples are given in [8]. Of course, not all structures lend themselves to composite manufacture, and it is therefore important to recognise which shapes and features make a part a suitable candidate for moulding (and which features do not!). The shape and features of a composite part can also have an influence on the selection of mould-making materials which in turn can affect the mould design and construction method. Other factors that need consideration include (but are not limited to) the speed and ease of construction, mould durability and fabrication cost.

6.2 Composite Part Design

A composite structure (or part) must be able to be extracted from the mould tool, usually without causing damage to either the composite or the mould [7]. In some instances, this requirement may mean that a designer needs to invest extra time to redesign some features. It is worth the time to reconsider the shape, functional needs and surface finish of the part prior to making a mould, as some simple design modifications can notably improve a composite part's mouldability [5]. To minimise moulding issues, composite structures should usually be designed to be as simple as possible.

Draft Angles. A draft angle defines the degree of taper on a moulded part. It is the angle between the sides of the part (or sides of the mould) and the extraction direction—zero, positive and negative drafts are shown in Fig. 6.1. A positive draft angle helps a part to be demoulded, whilst a negative draft angle (or undercut) restricts demoulding. Larger draft angles are usually required on deeper parts to prevent adhesion as the resin shrinks onto the surface during curing [9], whilst shallow parts can be adequately accommodated with relatively small drafts (1–3°) [10]. An appropriate draft angle minimises the friction between the composite and mould, and reduces mould wear (increasing mould life). Sensible draft angles reduce the likelihood of damage to the mould surface (and/or part) during the demoulding process.

As a rule of thumb, shallow parts (up to ≈75 mm in depth) should have a draft angle of 1° [1] (or more) for easy release. An extra degree of draft is recommended here for each additional ≈25 mm depth (or division thereof) [7], viz.

$$\theta \geq \lceil 1 + \frac{\text{depth [mm]} - 75}{25} \rceil \tag{6.1}$$

Of course, it should be noted that any draft angle (however small) is better for demoulding parts than no draft angle at all, but it is also important to be aware that reduced drafts are likely to cause difficulty during part removal and increase the chance of mould damage. Negative drafts or *undercuts* should be avoided whenever possible as these design features complicate part removal. For parts that must accommodate negative drafts, a *combination mould* [6] (also known as a *split* mould) can be used—see Fig. 6.2. Alternatively, removable inserts [5, 11] may be introduced at the mould design step (see Sect. 6.3). In a split mould, the interface where the two

6.2 Composite Part Design

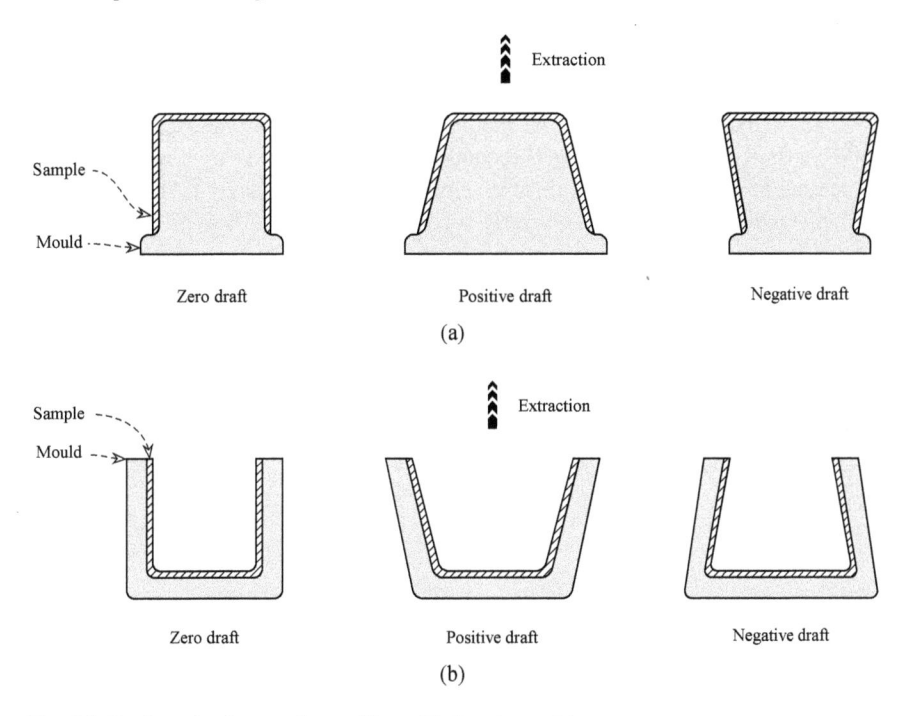

Fig. 6.1 Draft angles for **a** male moulds; and **b** female moulds

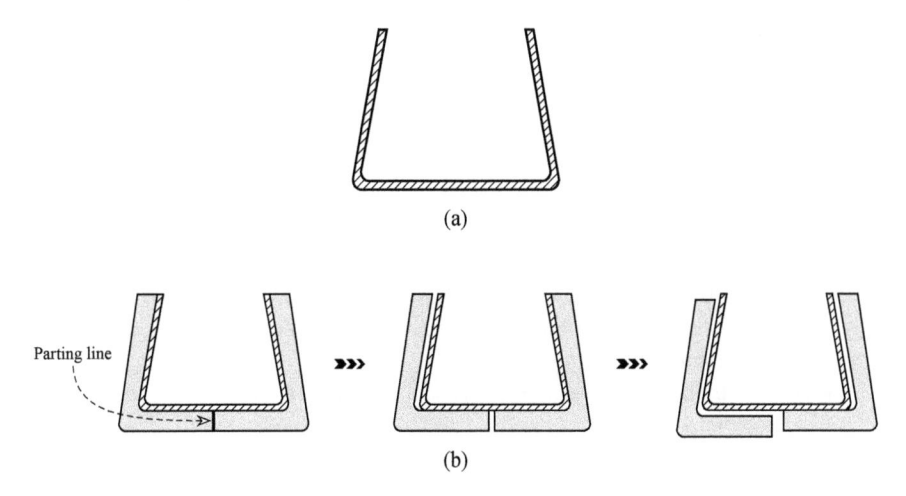

Fig. 6.2 Moulding a part with an undercut: **a** problematic part; and **b** combination mould

halves of the mould meet is referred to as the parting line; it is positioned to enable the mould tool halves to move independently of one another and hence, facilitates demoulding.

Example 6.1

A FRC channel section is to be fabricated in a female mould—see Fig. 6.3. What draft angle, θ, would you recommend for the channel to minimise any demoulding issues? What alternatives should be considered if the recommended draft angle cannot be easily accommodated in the design?

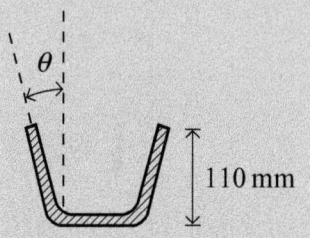

Fig. 6.3 Schematic profile of the channel

Solution

The draft angle (in degrees) should be

$$\theta \geq \lceil 1 + \frac{\text{depth [mm]} - 75}{25} \rceil$$

$$\theta \geq \lceil + \frac{110 - 75}{25} \rceil \geq \lceil 2.4° \rceil \geq 3°$$

If 3° cannot be accommodated,

- A redesign of the channel could be considered.
- It may be possible to mould with a reduced draft angle ($< 3°$). A reduction in the draft angle will, however, increases the friction between the channel and the mould. This will increases the likelihood of damaging the channel and/or mould tool during demoulding.
- A split mould tool (perhaps a vertical split at the mid-plane) could be considered.

Sharp Corners. Sharp corners or tight radii on a part are design features that can cause problems during composite moulding and should be avoided [1]. It is difficult to ensure fibres are tightly positioned against a mould when sharp discontinuities are present. Trying to create sharp corners on a part during the moulding process can lead to *bridging* [12]; fibres span the sudden contour change, rather than being firmly pressed against the mould surface—see Fig. 6.4a. As a bare minimum, a fillet of at least 5 mm is recommended here—$1/4$ inch (6.35 mm) is suggested in [13]. To

6.2 Composite Part Design

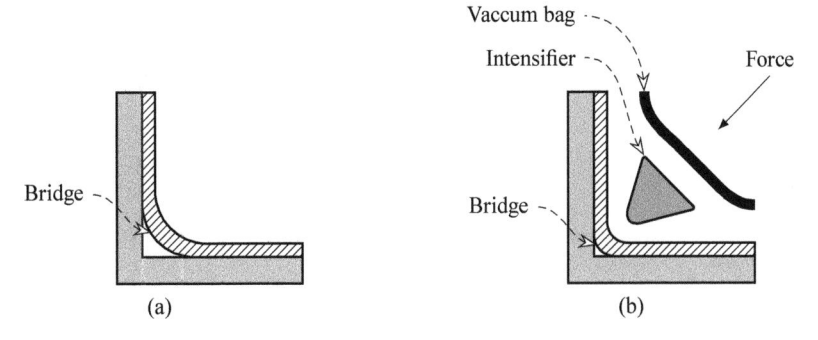

Fig. 6.4 Bridging at the corners of a mould: **a** bridge across a sharp corner; and **b** reduced bridge by using an intensifier

minimise bridging for composite parts with tight radii, pressure intensifiers can be used during the vacuum bagging process [2] as shown in Fig. 6.4b.

Surface Texture. A surface texture or pattern can (in many cases) be replicated on a composite component, but fine detail can sometimes be difficult to reproduce accurately. Thus, where possible, mirror-like finishes are recommended as parts can be released more easily from a smooth mould surface [5, 7]. Whilst surface patterns can camouflage imperfections, they offer an increase in surface area and (as a result) an increase in the chance for part adhesion during moulding. As mentioned, increasing the surface friction makes the release of a composite part from the mould more problematic so if rougher surface textures are unavoidable, consideration should be given to larger draft angles [11]. Peters [7] suggests an extra 1° of draft angle on vertical surfaces for each additional 0.0254 mm (0.001 inch) of texture depth.

6.3 Mould Design and Construction

Here, for simplicity, mould design and mould construction are combined in a single section.

The design and construction of a mould sometimes starts with a *plug* [14]; the terms *master* or *pattern* are used interchangeably [2, 5]. Of course, a mould can be manufactured directly, without first making a plug [1]. Direct mould manufacture is often employed when CNC machine tools are available, or for mould shapes with planar or simple curves. If a plug is to be used, its design is critical to the success of the mould and hence, to the subsequent fabrication of the composite part. A plug may be an original structure or part (as is often the case for the home fabricator), a model to be replicated, or as simple as a basic representation of the most important features. The mould is created around the plug and then used to make the finished composite part as shown in Fig. 6.5. In the figure, a male plug (i.e. the representation of the mouldable part) is used to construct a female mould and then finally the composite

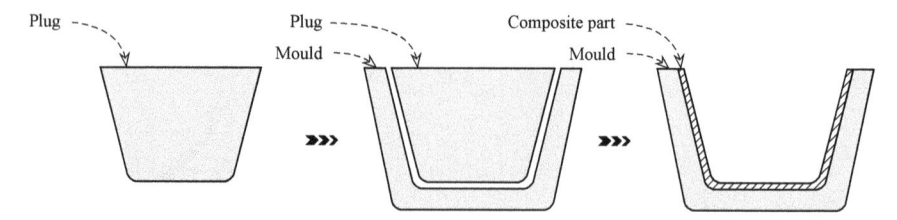

Fig. 6.5 Mould-making stages (from left to right): plug, creating the mould and moulding the composite part

part with a smooth convex surface. Since the plug exhibits the same features as the finished part, the plug must address the same design constraints previously discussed in Sect. 6.2.

In considering the plug and mould specifications, it should be evident that the surface finish of the plug and mould should be, at least, as good as the finish expected on the composite part [1, 4]. A plug is rarely used more than a few times, so its durability is often considered to be of lower importance. In contrast, a mould's lifespan is often the main design consideration; a mould may be used only a few times for prototyping or many hundreds (or even thousands) of times for high volume production parts [15].

Simple Low-Cost Moulds. Mould tools can be designed and constructed from easily accessible and low-cost materials such as plaster, foam and wood [4, 12, 16]; these materials can work well for many applications, but often need to be sealed [1, 17] and sanded to provide a smooth surface finish. Plaster, wood and foam all offer a relatively simple and quick fabrication route that requires only basic hand tools (i.e. wood and metal working tools [5]) to ensure successful mould construction. Nevertheless, they are not as durable as other tooling options, and they therefore tend to be used only for prototypes (perhaps 1–5 parts) or low volume production (5–50 parts) [15].

Silicone rubber offers an alternative to plaster, wood or foam. It is often preferred when there is a need to replicate complex or high (fine) detail, low-volume parts [2]. The use of a metal (or thin plastic) sheet formed around a substructure (often constructed from wood) is a further low-cost option used to increase mould durability. A metal sheet is non-porous and therefore does not need to be sealed post fabrication.

Additive manufacturing or 3D-printing typically uses thermoplastic polymers and is growing in popularity as a relatively low-cost option for mould-making [15, 18]. It offers many advantages for small parts, including the promise of tighter dimensional tolerances.

In considering mould-making materials, it is sometimes suggested that a mould should be 3–4 times (stiffer and) stronger than the resultant laminate [5]. Extra mould strength can obviously be achieved simply by increasing the thickness of the mould. In some instances, however, a better option is to support the mould with a network of braces [2]. This bracing method minimises the mass of the mould which can be beneficial where heat transfer to the mould (during the curing process) is important.

6.3 Mould Design and Construction

Example 6.2

A composite fabricator wants to construct a low-cost mould for prototyping a FRC flying disc (i.e. a Frisbee™). An existing *off-the-shelf* low-density polyethylene (LDPE) disc will be used as the plug—see Fig. 6.6. The fabricator will select one of the following mould-making materials for the prototype tooling:

a. Plaster.
b. Silicone.

Discuss the technical issues that might be foreseen for each of these materials?

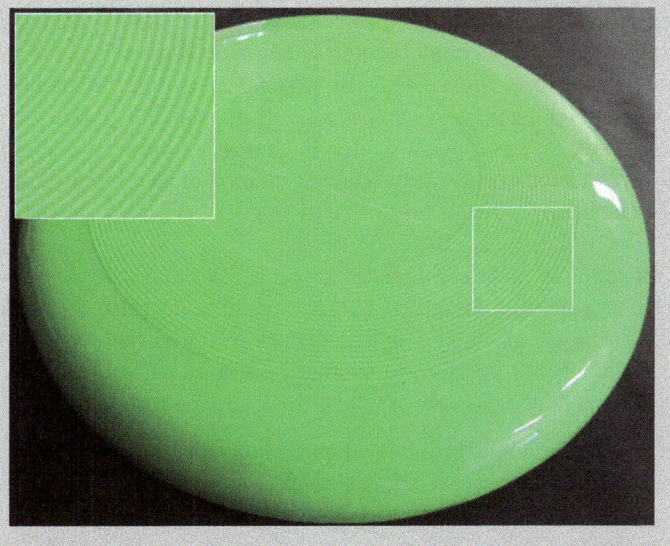

Fig. 6.6 An *off-the-shelf* plastic frisbee

Note. The issue of whether a FRC is the most appropriate material selection for a flying disc (frisbee) is not considered here. The flying disc is chosen on the basis that it offers some obvious moulding challenges for a composite part.

Solution

(a) Plaster

The frisbee has sensible draft angles (no undercuts) and is shallow without sharp corners or tight radii. Nevertheless, it still offers some clear mouldability concerns—fine detail is evident on the flying disc's surface in the form of a circular pattern, and the frisbee itself (although not evident just from the photo) may be too flexible for use as a plug without modification.

Fig. 6.7 Issues in the reproduction of fine detail using a plaster

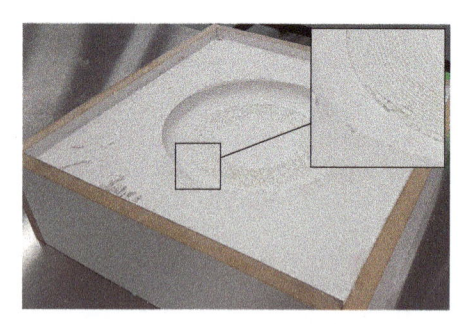

Surface textures or patterns that incorporate such fine detail can be problematic to replicate, an issue exacerbated by the use of some mould-making materials, including plaster. A plaster mould is not able to reproduce the frisbee's pattern—see Fig. 6.7. Thus, if a plaster mould is to be used, the frisbee design needs some modification—a smooth composite frisbee without the pattern is one option (see Sect. 6.4). This design change, however, will affect flight stability and may not be the most sensible solution in all circumstances—it is a design decision!

Note. The circular pattern introduces turbulence at the frisbee's leading edge which helps to stabilise the airflow over the top of the disc [19]. If it is essential to maintain the same flight stability characteristics as the original frisbee, consideration should be given to alternative design options [20] or to other mould-making materials, such as silicone that can replicate fine detail.

The *plug flexibility* must also be addressed before mould construction commences. The frisbee must be stiffened to ensure it retains its shape during the mould-making process. A simple way to do this is to simply fill the frisbee with plaster.

(b) Silicone

The use of a silicone mould enables fine detail to be accurately reproduced [2] and therefore removes the problem of replicating the plug's circular pattern. To mirror the pattern from the mould to the composite prototype, a gelcoat could be used to reproduce the fine detail and further provide a smooth surface coating for the flying disc—of course, in this case, the frisbee pattern is filled only with gelcoat and would contain no fibres. Moreover, it should be noted that gelcoats are usually brittle and, as a result, the surface would be prone to chipping.

6.3 Mould Design and Construction

Higher Volume Moulds. To produce more robust moulds, FRCs (*composite tooling*) or metals (*hard tooling*) are chosen. Composite tooling is often easier to fabricate than its metal counterpart and is a lower cost option that uses similar materials and fabrication methods as the laminates they are used to create. The cost and speed to build composite tooling is still higher than the aforementioned low-cost but less durable options of plaster, wood, foam and silicone. The use of composite tooling is usually suitable only for low to medium volume (in some cases up to several hundred) parts [14]. In contrast, hard tooling can be used for several hundred to several thousand composite parts [15] but it is heavy, more expensive and often needs specialist metal machining tools (e.g. CNC machine tools) for in-house fabrication. Hard tooling is therefore often out of reach for the small-scale composite fabricator.

Each of the mould-making materials available to the composite designer has its own advantages and disadvantages (or limitations), and there are many different approaches that will result in a successful outcome. The main advantages and disadvantages of each mould-making material, as well as recommended uses, are shown in Table 6.1.

6.4 Practical Task: Making a Frisbee

In this section, we expand on Example 6.2. A step-by-step guide to the design and manufacture of a prototype frisbee (including mould design and construction) is introduced for the plaster mould *only*—see Fig. 6.8a. A frisbee fabricated using a silicone mould (see Fig. 6.8b) was produced in a similar way, but a white gelcoat was used to create the surface pattern and to offer a smooth finish for the glass fibre substrate.

Design Specification. The aim is to produce a simple low-cost FRC frisbee that is easy to manufacture. A target mass of 175–200 g is identified for the frisbee [21] as well as an improvement in disc stiffness (elastic modulus) compared to the off-the-shelf LDPE disc (as shown in Fig. 6.6). Design improvements for the frisbee in terms of aerodynamics and flight stability are beyond the scope of this project, but if these issues are of interest, the reader is directed to reference [22] for further information.

Before the frisbee prototype can be fabricated (using our six-step process—see Fig. 3.1), a mould is needed.

Making the Mould. Here, a plaster mould is constructed. The off-the-shelf frisbee is used as the plug, but as previously mentioned, plaster will be unable to replicate the circular pattern on the frisbee's surface; hence, the pattern needs to be removed.

The pattern removal can be treated in one of the two ways:

- The pattern can be removed (smoothed) on the plug and the mould constructed from the smoothed frisbee.
- The mould can be formed with the pattern included on the plug and then the imperfections filled and sanded to produce a smooth mould surface.

Table 6.1 Advantages and limitations of mould-making materials [2, 5, 6, 12]

Material	Advantages	Disadvantages	Used for:
Plaster	• Easy to make. • Lowest cost option.	• Short lifespan (1–5 parts). • Needs drying and sealing. • Low temperature moulding (thermal shock). • Low (course) detail only.	• Plugs and one-off moulds. • Breakdown moulds (see Chap. 7). • Planar surfaces and surfaces with either simple or compound curves. • Low detail moulds.
Wood	• Easy to construct. • Low-cost.	• Relatively short lifespan. • Needs drying and sealing. • Low to moderate moulding temperatures only (heat distortion). • Complex shapes can be difficult to produce.	• Plugs, prototyping or low volume moulds (typically <50 parts), or bracing structures. • Planar surfaces and surfaces with simple curves.
Sheet metal	• Easy to cut and form.	• Performance depends on substructure (often wood). • Complex shapes are difficult to produce.	• Low to medium volumes. • Planar surfaces and surfaces with simple curves. • Low detail moulds.
Foam	• Easy to shape. • Low-cost. • Complex shapes (skill-dependent). • Low-medium temperature moulding.	• Relatively short lifespan—low density foams can be easily damaged. • Needs sealing and sanding. • Higher volumes are not usually possible.	• Plugs, one-off moulds or prototyping (usually 1–5 parts). • Planar surfaces and surfaces with simple or complex curves. • Low detail moulds.
Silicone Rubber	• Easy to make. • Low-cost. • Complex shapes and fine (high) detail. • Low to high moulding temperatures are possible.	• Relatively short lifespan (typically <50 parts). • High coefficient of thermal expansion (can also be a positive).	• Low volume moulds. • Planar surfaces and surfaces with simple or complex curves. • High (fine) detail moulding applications.
Composite Tooling	• Easy to make. • Cheaper and lower mass than hard tooling.	• Less durable than hard tooling. • Low to moderate moulding temperatures only (resin dependent).	• Moderate volume moulds (often >50 parts). • Planar surfaces and surfaces with simple or compound curves.
Hard Tooling	• Long lifespan. • Low to high temperature moulding. • Complex shapes and high (fine) detail are possible.	• Heavy (high density). • Expensive (highest cost). • Specialist machine tools are required. • High coefficient of thermal expansion (can also be a positive).	• High volume moulds (100–1000 parts or more). • Planar surfaces and surfaces with simple or compound curves. • High (fine) detail moulds.

6.4 Practical Task: Making a Frisbee

Fig. 6.8 Fibre-reinforced composite frisbee fabricated using a **a** plaster mould; and **b** silicone mould

Fig. 6.9 Smoothed frisbee plug (circular pattern removed)

It is usually better to resolve issues when making a plug, rather than later making significant amendments to a mould (or multiple moulds). Thus, smoothing the frisbee plug is preferred here. Sanding will smooth the surface but it will not result in the same high-gloss finish as the original frisbee—see Fig. 6.9. The *dull* sanded finish is adequate for a plaster mould as there will still be a need for some minor mould detailing, i.e. the plaster mould will need to be sealed and sanded after it has cured. For mould-making materials that require no sealing or sanding, the finish on the plug should be at least to the same standard as that needed on the finished composite part [5].

The *plug flexibility* must also be addressed before mould construction commences. To increase the frisbee stiffness and ensure it retains its shape during the mould-making process, the smoothed frisbee is filled with plaster; a plaster-based cement (cornice cement) is used here in preference to plaster of Paris. The cornice cement is mixed thoroughly and then poured into the frisbee base. It is left to cure as shown in Fig. 6.10. A *wet* (low viscosity) mix is used to help the plaster flow and fill the cavity.

Fig. 6.10 Plug stiffened with cornice cement

Fig. 6.11 Frisbee plug, finished and ready for mould creation

Next, the stiffened and smoothed frisbee is temporarily bonded to a glass plate—a gap of approximately 5 mm is created between the frisbee and plate. This gap is used to lift the disc off the glass, providing a deeper cavity in the mould that facilitates trimming and finish of the FRC frisbee after demoulding. Modelling clay is used to bridge the gap and remove the sharp discontinuity at the frisbee-plate interface as shown in Fig. 6.11.

The complete frisbee plug (stiffened frisbee and glass plate) is released with at least four coats of wax [3], and a wood structure is constructed around the glass plate to define the mould boundary—see Fig. 6.12a. To create the plaster mould, another batch of cornice cement is mixed to the consistency of honey and then slowly poured inside the wood boundary and over the stiffened frisbee—see Fig. 6.12b. The cornice cement is then left to cure and hence, form the mould. The setting time will depend upon the volume of the cement and the ambient temperature. Do not be tempted to rush this step when making a mould from plaster; the mould may take several days (or more) without intervention to set hard and allow the removal of the plug without causing damage to the mould; the goal is simply to produce a smooth *negative* of the frisbee plug.

Once the plaster has cured, the plug can be carefully extracted from the mould and the surrounding wood structure can be disassembled. The unsealed plaster mould is

6.4 Practical Task: Making a Frisbee

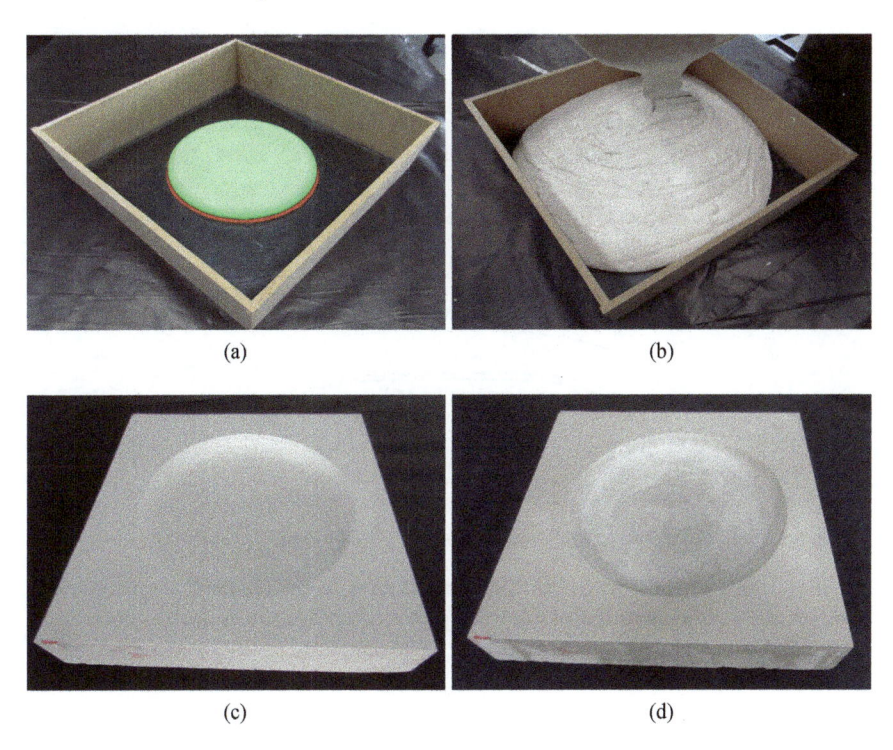

Fig. 6.12 Mould construction: **a** frisbee plug and wood 'boundary' structure; **b** pouring the cornice cement; **c** unsealed mould; and **d** sealed and sanded mould

shown in Fig. 6.12c, whilst Fig. 6.12d shows the mould after it has been sealed, sanded and polished. Here, two coats of low-viscosity (200–340 mPa · s) epoxy resin [23] are used to seal the mould. The epoxy is left to fully cure before the surface is carefully sanded to a smooth finish.

We now have the plaster mould, so we can make the frisbee prototype.

Frisbee Prototype. The frisbee is manufactured using wet layup with glass fibres (chopped strand random mat, 450 g/m²) in a polyester matrix [24]. These material selections offer the simplest and lowest cost options for a FRC frisbee—herein, the focus is on satisfying the low-cost requirement but it is recognised that other alternative material selections are just as valid. The number of plies are chosen to satisfy the mass and stiffness criteria—see Example 6.3. Two random mat plies (low V_f) are used to create the main structure of the flying disc, and a thickened rim is replicated with three extra random mat plies—note, a heavier rim is known to improve flight stability [22].

Example 6.3

How many plies of glass fibre random mat ($A_w = 450$ g/m^2) are needed to manufacture a FRC frisbee with a target mass of 175–200 g, and an elastic modulus in excess of a *typical* LDPE flying disc. Assume the reinforced rim has, at least, twice as many plies as the main structure of the flying disc—see Fig. 6.13.

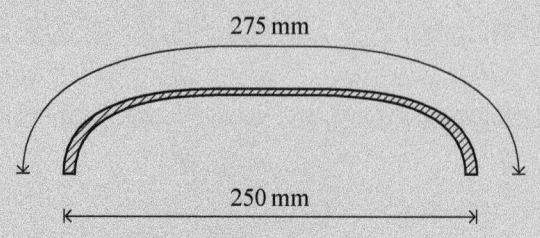

Fig. 6.13 Schematic cross-section of the frisbee

Note. The frisbee surface will need a decorative finish. To meet the frisbee's target mass, consideration of the weight of any surface finish is also required.

Solution

The range of fibre volume/weight fraction for wet layup is obtained from Table 3.2, i.e. $V_f = 0.1 - 0.3$ and $W_f = 0.14 - 0.48$. Based on fabrication experience from earlier frisbee prototyping activities using the same constituents, volume fraction and weight fractions at the medium to high range are targeted here. As a first step, we therefore assume $V_f = 0.25$ and $W_f = 0.41$; for fabricators with less experience, consideration should be given to lower fibre volume and weight fractions.

Mass. The target mass of the frisbee is 175–200 g. Therefore, to allow for a painted surface finish, an unpainted composite mass of approximately 175 g is considered.

Based on a uniform thickness, we can calculate the mass of the *main structure*. For a glass fibre random mat composite ($W_f = 0.41$), the mass can be calculated as

$$\text{Mass of fibre (per ply)} = A_w \frac{\pi D^2}{4} = 450 \, \frac{\text{g}}{\text{m}^2} \times \frac{\pi \, 0.275^2}{4} = 26.7 \, \text{g}$$

$$\text{Mass of resin (per ply)} = 26.7 \, \text{g} \times \frac{0.59}{0.41} = 38.4 \, \text{g}$$

6.4 Practical Task: Making a Frisbee

$$\text{Mass of each ply} = 26.7 + 38.4 = 65.1\,\text{g}$$
$$\text{Mass for 2 plies} = 2 \times 65.1 = 130.2\,\text{g}$$

Assuming two plies, the thickness of the main structure is calculated from

$$t = \frac{nA_w}{\rho_f V_f}$$
$$= \frac{2 \times 0.450}{2550 \times 0.25} = 1.4\,\text{mm (i.e. 0.7 mm per ply)}$$

Now, considering the extra *rim reinforcement* (width = 20 mm)

$$\text{Mass of extra fibre (per ply)} = A_w \times \pi D_{\text{rim}} \times \text{width}$$
$$= 450\,\frac{\text{g}}{\text{m}^2} \times \pi 0.25 \times 0.02 = 7.1\,\text{g}$$
$$\text{Mass of resin (per ply)} = 7.1\,\text{g} \times \frac{0.59}{0.41} = 10.2\,\text{g}$$
$$\text{Mass of each rim ply} = 7.1 + 10.2 = 17.3\,\text{g}$$
$$\text{Mass for 3 plies} = 3 \times 17.3 = 51.9\,\text{g}$$

Total frisbee mass = mass of main structure + mass of rim reinforcement
$$= 130.2 + 51.9 = 182.1\,\text{g}$$

Thus, the frisbee mass prediction should be within the target mass range (based on the design specification), assuming not more than 17.9 g is used for finishing (painting).

Stiffness. The elastic modulus of the composite frisbee must be greater than that of an *off-the-shelf* LDPE frisbee.

The elastic modulus of the glass fibres and polyester are assumed to be 70 GPa and 3 GPa, respectively, since no supplier property data was readily available for either constituent.

Thus,

$$E_c = E_m(1 - V_f) + E_f V_f$$
$$= 3(1 - 0.25) + \frac{3}{8} \times 70 \times 0.25 = 8.8\,\text{GPa} \gg E_{\text{LDPE}}(\,0.172\text{--}0.282\,\text{GPa}\,[27])$$

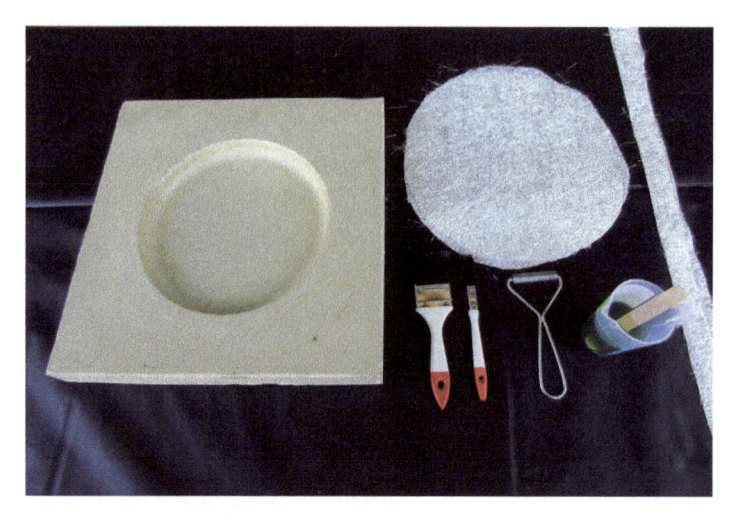

Fig. 6.14 Materials and equipment for wet layup

> The stiffness prediction therefore satisfies the design specification, i.e. $E_c > E_{\text{LDPE}}$.
>
> **Summary**
> Based on the initial design (sizing) calculations:
>
> - Two plies (glass fibre random mat) should be used for the main structure of the frisbee (1.4 mm thick).
> - An extra three random mat plies should be used to thicken the disc's rim (an extra 2.1 mm).
> - To meet the frisbee target mass (175–200 g), the painted coating should have a mass not more than 17.9 g.

Prior to layup, at least 4 coats of wax release agent should be applied to the surface of a new mould [3]. Here, six coats are used as, from previous experience, demoulding parts from plaster moulds can prove troublesome. The wax is buffed to a sheen between coats in accordance with the supplier's instructions.

After waxing is complete, the layup and consolidation step commences. All materials and equipment are initially set up in a well-ventilated area—see Fig. 6.14. The fibres are precut to the requisite size, and the resin is premixed with 1.5% catalyst (in accordance with the supplier's instructions) assuming a wastage factor of 1.25—see Example 6.4. Here, the fibres are cut oversize and will be trimmed back later after demoulding.

6.4 Practical Task: Making a Frisbee

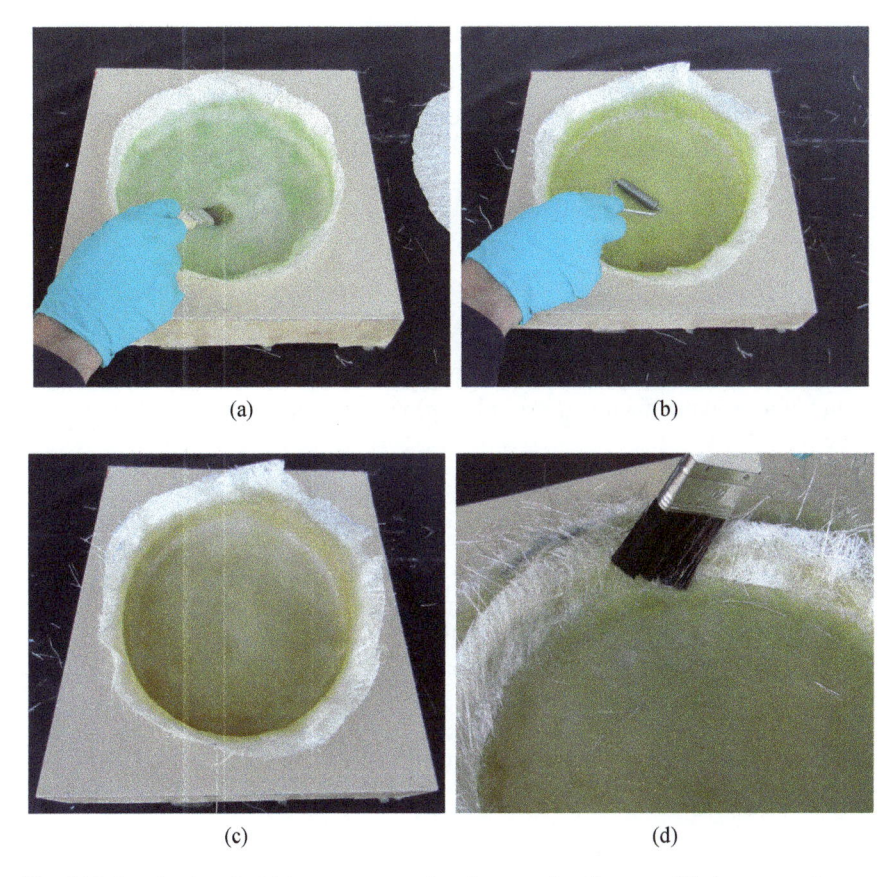

Fig. 6.15 Lamination of a frisbee: **a** wet out of random mat; **b** roller consolidation; **c** complete wet out; and **d** rim reinforcement

To create the frisbee, a random mat ply is positioned in the mould and the polyester is applied with a brush to wet out the random fibres as shown in Fig. 6.15a. The random mat is worked with a stippling action to ensure the fibres are pushed flat against the mould surface. This is repeated for the second layer of fibres. After each ply, a roller is used to remove air trapped between the random fibres as shown in Fig. 6.15b. The wet-out and rolled layup is shown in Fig. 6.15c. Lastly, the extra plies are added to reinforce the disc's rim (see Fig. 6.15d)[1]. This is a tricky task to perform at this stage so if necessary, the rim reinforcement can be added after the two plies (for the main structure) have cured.

---

[1] Note. The image in Fig. 6.15d is not from the same frisbee fabrication process as Fig. 6.15a, 6.15b and 6.15c.

Example 6.4

Consider Example 6.3. Assuming the resin is to be premixed with 1.5% catalyst using a wastage factor of 1.25, how much polyester resin and catalyst are needed?

Solution

Based on previous experience with the glass fibres and polyester, a medium-high fibre volume fraction (for wet layup) was chosen in the earlier example, i.e. $V_f = 0.25$.

Thus, $W_f = 0.41$ and hence

$$\text{Resin and catalyst mix (g)} = 182.1 \times (1 - 0.41) = 107.4 \, \text{g}$$

To take account for wastage, a factor of 1.25 is applied, viz.

$$\text{Resin and catalyst mix (g)} = 107.4 \times 1.25 \approx 134.3 \, \text{g}$$

The polyester resin used here is mixed with only 1.5% catalyst, i.e. $^{98.5}/_{100}$ of resin to $^{1.5}/_{100}$ of catalyst. Thus

$$\text{Resin (g)} = 134.3 \times 0.985 = 132.3 \, \text{g}$$
$$\text{Catalyst (g)} = 134.3 \times 0.015 = 2.0 \, \text{g}$$

The frisbee is left for several hours until it reaches a solid cure before it is removed from the mould. In some instances, demoulding can be performed using a plastic wedge to tease the disc's rim away from the mould, allowing the plaster mould to be reused. Of course, a notable benefit of using a plaster mould is that (in extreme cases) it can be broken to release the composite part—others refer to this as a *breakdown* method [12]. The demoulded frisbee is shown in Fig. 6.16.

To finish the frisbee, the disc's rim is trimmed and lightly sanded to create an even rounded edge—see Fig. 6.17a. The disc's surface is sanded, cleaned and finally painted. Sanding commences with using 220 grit and then continues to progress to finer grades (i.e. P220–≈P600). A primer is used to ensure maximum adhesion and to remove any minor surface imperfections that may still remain. The frisbee is finished with a black acrylic paint.

6.4 Practical Task: Making a Frisbee

Fig. 6.16 Demoulded FRC frisbee with excess fibres

Fig. 6.17 Frisbee post-processing: **a** trimmed; and **b** sanded and painted

Physical Testing. The final mass of the composite frisbee is measured at 182.0 g. This is not too dissimilar to the estimated unpainted mass of 182.1 g—see Example 6.3. By flexing the composite flying disc, it is clear that the prototype is much stiffer than the off-the-shelf LDPE frisbee that was used to make the plaster mould. So, now all that is left to do is to test the flight characteristics of our FRC prototype!

6.5 Summary

To create useful composite structures or parts, consideration should be given to part design in the context of mouldability as well as to mould design and construction issues.

In terms of *composite part design*, the following questions (as a minimum) should be reviewed:

- Are the draft angles appropriate?
- Are there negative drafts (undercuts)?
- Are there any surface texture issues?
- Are there any sharp corners or tight radii?

For *mould design and construction*, the following questions are mentionable:

- Is there a need for a plug? If *yes*, will an existing part be used, or does the plug need to be fabricated?
- How many parts will be moulded?
- How accurate does the finished laminate (and hence, mould) need to be? If there are tight tolerances, specialist machine tools may be needed for mould construction.
- How much time is available to build the mould tool, and what is the budget?

When considering *mould-making materials*, the best selection depends on many factors including the speed and ease of manufacture; surface finish requirements; mould durability; and fabrication cost.

Mould-making materials for simple low-cost tooling include:

- Plaster.
- Wood.
- Foam.
- Silicone.

For higher volume moulds, materials typically used are:

- Composite tooling.
- Hard (metal) tooling.

6.6 Questions

Question 6.1 What is a *draft angle* and why is it important in moulding FRCs?

Question 6.2 What is the minimum (recommended) draft angle for shallow composite parts? For deeper parts, what should the draft angle be?

6.6 Questions

Question 6.3 Why is it important to avoid sharp corners on composite parts? What minimum fillet radius would you recommend?

Question 6.4 What is *bridging* and why should it be avoided?

Question 6.5 Why should surface textures on composite parts be avoided?

Question 6.6 What is a *plug*? What other terms are used interchangeably to mean plug?

Question 6.7 Why is the durability of a plug usually of lesser importance than that of a mould?

Question 6.8 List four low-cost mould-making materials and two materials that might be used for higher volume moulds.

6.7 Problems

Problem 6.1 Which mould in Fig. 6.18 is *male*, which is *female* and which is a *matched mould*?

Answer. a. Female; b. male; c. matched mould.

Problem 6.2 Which of the moulds in Fig. 6.19 are likely to allow the composite part to be demoulded from the tool? If demoulding is possible, comment on how difficult this might be.

Problem 6.3 The cross-section of a thin-walled composite part and its mould tool are shown in Fig. 6.20. Identify issues that might arise when moulding the part and then make suggestions (in terms of part and/or mould design) to ensure a successful composite project.

Answer. Issues include mould stiffness(?), sharp corners, no drafts on some sections and an undercut (negative draft).

Problem 6.4 Identify the most appropriate mould-making material based on the moulding requirements in Fig. 6.21.

Answer. Top-to-bottom: sheet metal (wood substructure); metal; and composite.

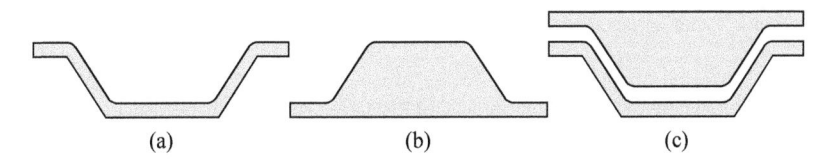

Fig. 6.18 Male, female and matched moulds (mould types to be identified)

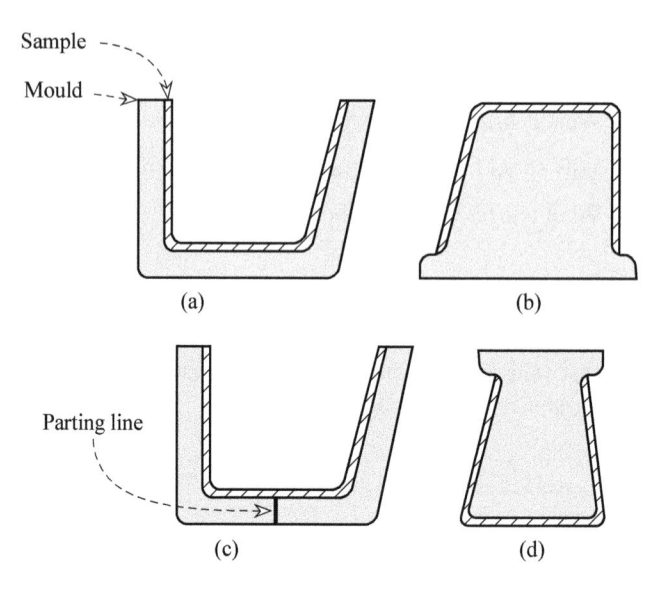

Fig. 6.19 Demoulding (demoulding issues, if any, to be identified)

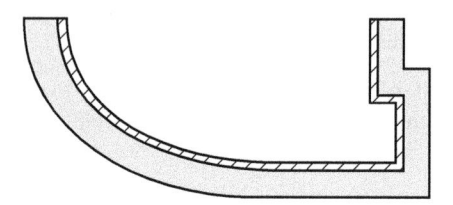

Fig. 6.20 A composite part and its mould tool (moulding issues to be identified)

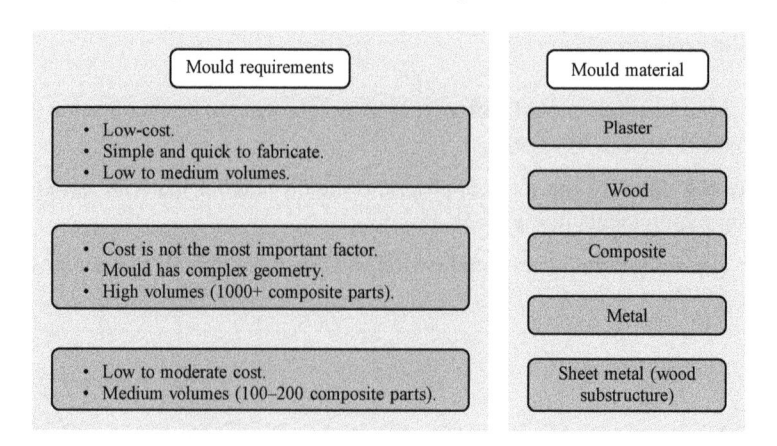

Fig. 6.21 Mould requirements and material selection (materials to be identified)

References

1. Astrom BT (2018) Manufacturing of polymer composites, 2nd edn. Routledge, Boca Raton
2. Strong AB (2008) Fundamentals of composites manufacturing: materials, methods and applications, 2nd edn. Society of Manufacturing Engineers, Dearborn, Mich
3. Wanberg J (2009) Composite materials: fabrication handbook #1, composite garage series, vol 1. Wolfgang Publications, Stillwater, Minnesota
4. Fibreglast (2019) Mold construction guide. https://www.fibreglast.com/product/mold-construction/LearningCenter
5. Wanberg J (2010) Composite materials: fabrication handbook #2. Composite garage series. Wolfgang Publications, Stillwater, Minnesota
6. Wang R, Zheng S, Zheng Y (2011) Polymer matrix composites and technology. Woodhead publishing in materials, Woodhead Publication and Science Press, Oxford and Philadelphia and Beijing
7. Peters ST (1998) Handbook of composites, 2nd edn. Chapman & Hall, London
8. Wanberg J (2012) Composite materials: fabrication handbook #3, composite garage series, vol 3. Wolfgang Publications, Stillwater, Minnesota
9. Biron M (2004) Thermosets and composites: technical information for plastics users/Michel Biron. Elsevier, Oxford
10. Mallick PK (1997) Composites engineering handbook, materials engineering, vol 11. Marcel Dekker, New York
11. Carlsson LA, Gillespie JW (1989-91) Delaware composites design encyclopedia. Technomic, Lancaster
12. Lee SM (1992) Handbook of composite reinforcements. VCH, New York
13. Aird F (2014) Fiberglass and other composite materialshp1498: a guide to high performance non-metallic materials for automotiveracing and marine use. includes fiberglass, kevlar, carbon fiber, molds, structures and materials. HP Books, New York
14. Pauer R, Pokelwaldt A (2017) Composite molds: choices and considerations. http://compositesmanufacturingmagazine.com/2017/11/best-practices-for-choosing-composite-molds/
15. CompositesWorld (2016) Tooling. https://www.compositesworld.com/articles/tooling
16. Barbero EJ (2017) Introduction to composite materials design, 3rd edn. Composite materials. CRC Press, Boca Raton
17. Weatherhead RG (1980) FRP technology: fibre reinforced resin systems. Springer, Netherlands
18. CompositesWorld (2015) A growing trend: 3d printing of aerospace tooling. https://www.compositesworld.com/articles/a-growing-trend-3d-printing-of-aerospace-tooling
19. Motoyama E (2002) The physics of flying discs. URL: https://www.people.csail.mit.edu/jrennie/discgolf/physics.pdf
20. Polar Manufacturing (2020) Inspiration. www.polar-manufacturing.com/inspiration/
21. Carter J (2020) 10 best frisbees in 2020. www.gearhungry.com/best-frisbee/
22. Potts J, Crowther W (2002) Frisbee(tm) aerodynamics. In: 20th AIAA applied aerodynamics conference, American Institute of Aeronautics and Astronautics, Reston. https://doi.org/10.2514/6.2002-3150
23. Kinetix (2019) Laminating: R118 infusion. http://atlcomposites.com.au/icart/products/14/images/main/KINETIXR118&H115-H120-H126-H103.pdf
24. Protite (2019) Fibreglass resin. https://www.protite.com.au/products/fibreglass/fibreglass-resin/
25. Callister WD, Rethwisch DG (2018) Materials science and engineering: an introduction, 10th edn. Wiley, Hoboken

Chapter 7
Hollow Sections—How to Make Composite Tubes

Abstract This chapter considers the manufacture of hollow sections. Mandrel lamination (wrapping) and bladder moulding are described and then used to create composite tubes. A step-by-step guide to tube manufacture is presented for prepreg moulding, but the fabrication methods introduced are more widely applicable (and hence, can be adapted for use in wet layup processes). Whilst the focus is on cylindrical tubes, the fabrication process can be simply modified to create square or rectangular hollow sections or more complex tubular structures. A helical spring and a bicycle handlebar are used as more complex examples. Demoulding methods for mandrel wrapped tubes are discussed. A specific focus is given to mechanical extraction and thermal (heating and cooling) mechanisms. The concept of mouldless composite construction is introduced. The importance of a mandrel's coefficient of thermal expansion is considered in the context of demoulding using thermal methods.

7.1 Introduction

The design and manufacture of composite laminates has been considered in earlier chapters. A composite can have a high stiffness and strength at fibre orientations in the plane of the plies, but tends to offer minimal resistance to out-of-plane loads [1]. To improve the flexural and torsional performance of a composite laminate without adding unnecessary weight, the laminate can be effectively 'thickened' by using a hollow section.

Hollow steel sections are widely used in structural applications. The volume of material is located at the outer extremes of the cross-section and this maximises the section properties [2], improving bending and torsional performance per unit mass. The manufacture of composite hollow structures offers weight benefits over those from traditional materials, and hence enables fabrication of even lighter and more economical structures.

There are numerous methods for manufacturing fibre-reinforced composite (FRC) hollow sections, but two simple methods are mandrel lamination (wrapping) [3] and bladder moulding [4]. In this chapter, a step-by-step approach for creating a cylindrical tube is presented for these two simple methods using prepreg. The approach is similar for wet layup. The methods can be easily adapted for square or rectangular cross-sections, or complex hollow shapes with cross-sectional properties that vary along the length of the composite. To illustrate this, the chapter finishes with two more complex design and manufacture examples: a helical spring and a bicycle handlebar.

7.2 Mandrel Lamination

In mandrel lamination, the fibres are wrapped around a *mandrel* [3], as shown in Fig. 7.1. A wet layup method [3] or prepreg moulding [1, 5] may be used. If wet layup is preferred, the resin can be introduced to the fibres either prior to or whilst wrapping each layer of fibre. The mandrel acts as the mould tool to create a smooth inner surface on the tube, similar to a male mould [6]. To create open-ended parts, a mandrel is often simply a cylindrical or tapered (conical) mould tool constructed from steel or aluminium [1, 7]. The mandrel defines the internal tube diameter and surface finish, but the external features of the tube are not precisely controlled [8]. The open-ends facilitate the removal of the finished (cured) composite from the mandrel [9].

Mandrel lamination is more complicated for closed-ended parts [1], or for parts with zero or negative drafts. Herein, we will consider tubes with no draft angle. For closed-ended parts, the simplest method is to leave the mandrel (often made of foam) inside the part after curing. If mandrel removal is needed, one method is to use a dissolvable mandrel [7]; sometimes, this is referred to as *mouldless* composite construction [10]. Polystyrene foam is compatible with epoxy resins (but not polyester) and can be dissolved with acetone [3] whilst wax mandrels (candle or paraffin wax) can be melted at elevated temperatures (ranging 50–90 °C). A low-cost and water-soluble mandrel can be easily created by combining sand and polyvinyl alcohol [11], or alternatively breakdown materials such as plaster can be used [7].

To demonstrate the mandrel lamination process, a cylindrical aluminium (6061) mandrel is used to create a straight hollow tube comprising four layers of 200 $^g/_{m^2}$ carbon twill weave prepreg; of course, a similar lamination outcome is possible with a wet layup process or by using other fibre types and forms. It should be noted that

Fig. 7.1 Mandrel lamination for an open-ended hollow tube

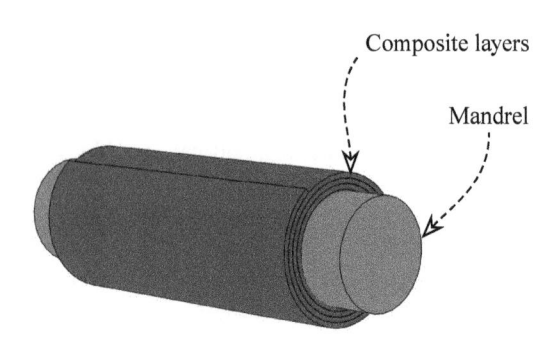

7.2 Mandrel Lamination

Fig. 7.2 The aluminium mandrel: 25mm OD with a smooth (mirror) finish

the six steps to design and make a FRC (see Fig. 3.1) are once again followed; the steps are not explicitly referenced herein, as it should (by now) be trivial to relate the specific fabrication detail to the relevant step in the composite manufacturing process.

The mandrel has an outer diameter of 25 mm and the surface is polished to a *mirror* finish—see Fig. 7.2. The warp fibres in each ply are aligned with the mandrel axis, whilst the weft fibres are orientated circumferentially around the tube.

For each of the four plies, a rectangle 250mm long (warp) and 90mm wide (weft) is cut from the roll of prepreg. The rectangle's width is slightly greater than the circumference of the mandrel (approximately 10mm extra) to allow one ply of fibres to exceed one complete rotation around the mandrel circumference. This method is preferred here but alternative processes can also be used, depending on the fabricator's preference; for example, a *Swiss roll* layup [12] or a multi-layer wrapping process (see Sect. 7.3).

To prepare the mandrel, it is first cleaned with warm water and a lint-free cloth, then left to dry. At least four layers of high-temperature wax (suitable up to 120 °C) are applied to the mandrel. Between each layer, the wax is carefully buffed to an even sheen in accordance with the supplier's instruction.

During layup, the prepreg is carefully wrapped around the mandrel—see Fig. 7.3. Tension is applied to the prepreg during roll wrapping to prevent creases. Here, the wrapping process focuses on one layer at a time. After the first ply is complete and smoothed against the mandrel's surface, the second, third and finally the fourth plies are wrapped. The start of each ply commences close to the finishing position of the previous one.

In the earlier flat plate demonstration of prepreg moulding (see Chap. 4), the composite was cured in an oven and consolidation was performed using a vacuum bag. Here, an opportunity is taken to introduce an alternative consolidation method using a *heat shrink tube* [13]—normally, this method of consolidation is used with wet resin. The shrink tube is placed over the fibres and mandrel (as shown in Fig. 7.4a), and is then either placed in the oven to shrink onto the tube during curing, or gently and uniformly heated (prior to curing) with a heat gun. The shrink tube should have an appropriate contraction temperature to ensure adequate consolidation as composite performance and surface finish are significantly influenced by consolidation pressure [14]. Here, the shrink tube and prepreg are simply placed directly in the

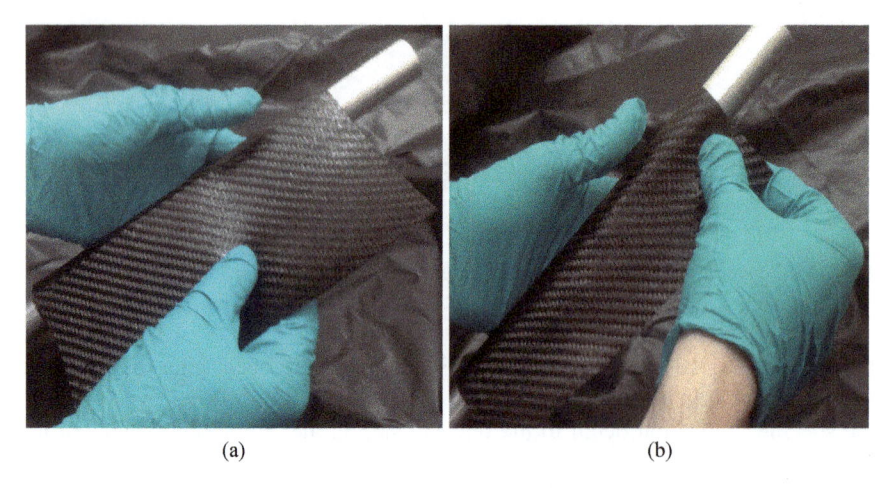

Fig. 7.3 Mandrel lamination: **a** prepreg tension; and **b** wrapping process

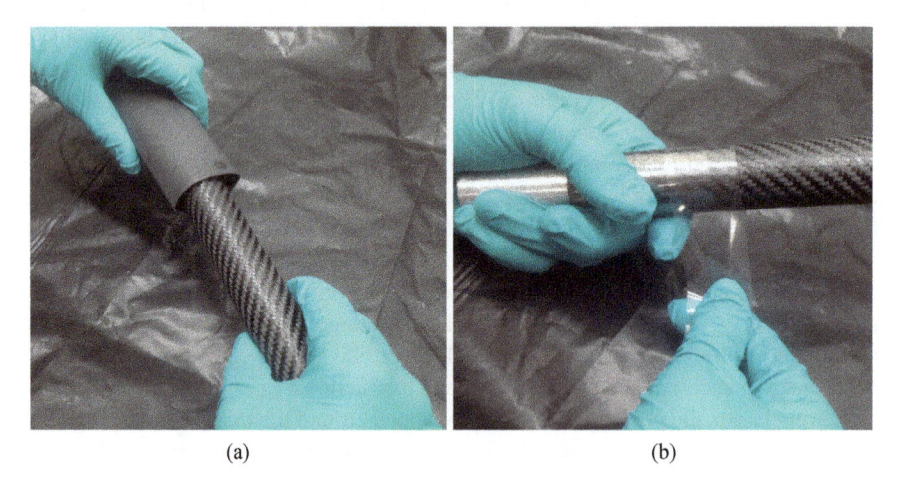

Fig. 7.4 Consolidation using **a** heat shrink tube; and **b** shrink tape

oven, allowing the shrink tube to consolidate the prepreg during curing at a dwell temperature of 120 °C.

A similar alternative to shrink tube is a *shrink tape* [15]. Shrink tape works in an equivalent manner to its tubular counterpart but is wrapped in a spiral manner around the preform, rather than slid over the top of the fibres and mandrel. Each spiral wrap of shrink tape should overlap the previous ones by at least half of the tape width [3] as shown in Fig. 7.4b.

Here, the composite tube is left to cure on the mandrel and the shrink tube is carefully cut and removed. A post-cure is performed before the composite part is extracted (demoulded) from the mandrel.

7.2 Mandrel Lamination

If some encouragement is needed during demoulding, three methods can be used to assist mandrel removal from an open-ended part:

- Mechanical extraction [3].
- Thermal (heating and cooling) methods [12].
- Compressed air-assisted demoulding [16].

Mechanical Extraction. Demoulding is often achieved by mechanical extraction. It is particularly common when wet layup is used. A constant force is applied to the mandrel (often using a press [3]) whilst the tube is suitably restrained (or vice versa). An alternative method is to introduce a *gentle* tap (dynamic force) with a mallet. The amount of mechanical force needed will depend upon the interference fit (between the FRC and mandrel) that results from the resin shrinkage during the curing process [11, 17].

Thermal (Heating and Cooling) Methods. Heating causes an aluminium mandrel to expand and cooling results in contraction. The amount of expansion or contraction is identified by the coefficient of thermal expansion (CTE). If a mandrel has a higher CTE than the composite, it will expand more when heated and contract more when cooled. This can be used to release the interference fit caused by the shrinking resin [12].

For a unidirectional FRC, the axial (longitudinal) CTE (α_{cl}) is given by [11]

$$\alpha_{cl} = \frac{\alpha_m E_m (1 - V_f) + \alpha_f E_f V_f}{E_m (1 - V_f) + E_f V_f} \qquad (7.1)$$

where α_m and α_f are the coefficients of thermal expansion of the matrix and fibres, respectively, E_m and E_f are the elastic moduli values, and V_f is the fibre volume fraction.

The transverse CTE (α_{ct}) for a unidirectional composite is given by [11]

$$\alpha_{ct} = \alpha_m (1 + \nu_m)(1 - V_f) + \alpha_f V_f \qquad (7.2)$$

where ν_m is Poisson's ratio of the matrix.

The fibres in a woven and random fibre composite are not all orientated in one direction. Woven fabrics have fibres orientated in both the warp and weft directions and hence, the CTE of the composite in each of these orientations will be similar, and will lie between the longitudinal and transverse threshold coefficients. Typically, the CTE is much closer to the longitudinal values [18]. The coefficients for random FRCs will be similar in all in-plane orientations and will also lie between the threshold values.

Compressed Air-Assisted Demoulding. The use of compressed air can be a simple and effective method for demoulding a composite part. A hole (or a series of holes) is drilled in the mandrel so that low-pressure air can be blown between the surface of the mandrel and the composite. This air pressure helps to release the composite from the mandrel.

In our mandrel lamination demonstration, the prepreg (and hence, mandrel) is heated during the curing process and then cooled to room temperature prior to demoulding. The elevated cure temperature for prepreg therefore effectively introducing a thermal assisted demoulding step as part of the usual manufacturing process. Thus, minimal force is needed to break any remaining seal and the tube simply slides off the mandrel despite the zero draft angle. However, if the curing process was performed at room temperature (as is the case for wet layup—see Chap. 3), the thermal demoulding method would require a reduction of temperature to shrink the mandrel from the FRC. In this instance, the composite can be placed in cold storage ($-18\,°C$) or introduced to liquid nitrogen (about $-200\,°C$) [19].

Example 7.1

Mandrel lamination is carried out with woven carbon fibre-reinforced epoxy prepreg on an aluminium mandrel. Assume α_c is about 4 μm/m°C [18] for the composite (i.e. for both the longitudinal and transverse orientations), and that of the aluminium mandrel α_{man} is 23.6 μm/m°C [20]. If the prepreg is cured at $120\,°C$ and then cooled to a room temperature of $20\,°C$ prior to demoulding, calculate the clearance between the mandrel and the tube. Assume no adhesion occurs and neglect the shrinkage of the epoxy during curing.

[**Note.** This is an estimation for the carbon fibre tube we are fabricating here.]

Solution 7.1

During the curing process, the mandrel and composite are heated up from room temperature ($20\,°C$) to $120\,°C$, and then cooled back down to $20\,°C$. The diameter of the mandrel at room temperature is $d_{man,20}$. The diameter expands as the temperature increases to $d_{man,120}$ and finally returns to its original diameter $d_{man,20}$ when cooled back to room temperature.

During heating, it is assumed that the FRC can expand freely since it has not yet reached a solid cure. In other words, the composite diameter $d_{c,120}$ will follow that of the mandrel.

Thus, the expansion of the mandrel diameter is

$$\Delta d_{man} = d_{man,20} \alpha_a \Delta t$$
$$= 25 \times 10^{-3} \times 23.6 \times 10^{-6} \times (120 - 20)$$
$$= 59 \times 10^{-6} \text{ m} = 0.059 \text{ mm}$$

7.2 Mandrel Lamination

and hence, the composite diameter after it is fully cured is

$$\rightarrow \quad d_{c,120} = d_{\text{man},120} = d_{man,20} + \Delta d_{\text{man}} = 25.059 \text{ mm}$$

During the cooling phase, the mandrel returns to its original diameter of 25 mm, whilst the composite tube undergoes a contraction related to its coefficient of thermal expansion, α_c,

$$\Delta d_c = d_{c,120} \alpha_c \Delta t$$
$$= 25.059 \times 10^{-3} \times 4 \times 10^{-6} \times (20 - 120)$$
$$= -10.0236 \times 10^{-6} \text{ m} = -0.0100236 \text{ mm}$$
$$\rightarrow \quad d_{c,20} = d_{c,120} + \Delta d_c = 25.059 - 0.0100236 = 25.049 \text{ mm}$$

The final diameter of the FRC after the curing process is $d_{c,20} = 25.049$ mm.

The clearance between the mandrel and composite will therefore be 0.049 mm.

To produce a smooth surface finish on the outside of a tube fabricated using mandrel lamination requires secondary processing tasks. Here, the ends of the composite tube are carefully trimmed using a band saw. The outer surface is then sanded, moving through a series of grits until greater than 400 grit, sanding finished at P800 (i.e. 400–500 grit). The surfaces are cleaned with a solvent (acetone) and rinsed with water then left to air dry before a clear spray paint (clear coat) is used as a decorative finish. Several layers of clear coat are applied in thin coats to produce a smooth and glossy finish. The carbon tube is shown in its near-finished state in Fig. 7.5—only a few minor surface imperfections remain.

Fig. 7.5 Carbon fibre-reinforced tube manufactured using mandrel lamination

7.3 Bladder Moulding

Bladder moulding [4] uses a cavity mould tool and an inflation *bladder*—see Fig. 7.6. Silicone or latex are commonly used elastomeric bladders [21], but a nylon film sheet, cut to shape and bonded on its edges has been used as a low-cost option [1, 4]. During moulding, the composite layup is positioned between the bladder and mould cavity. The bladder is internally pressurised and expands to fill the void and the composite is consolidated against the inner surface of the mould. The cavity therefore defines the external shape and surface finish of the hollow structure. To remove the bladder, the inflation pressure is simply reduced and the bladder returns to its undeformed shape.

Here, a hollow tube is created using a split (two-part) aluminium mould (6061)—see Fig. 7.7a. The internal mould cavity is used to create an external tube diameter of about 27 mm. Each of the split mould halves has four alignment holes. Moreover, there are four threaded holes in the upper half of the mould and four blind recesses in the lower half (at matching locations). The threaded holes and the matching recesses are to be used in demoulding and will be revisited later. To ensure the mould halves are correctly located, alignment pins are inserted into the alignment holes during the layup and the moulding process. The pins retain the two mould halves in place, ensuring their cavities align. A silicone tube, knotted at one end, is used as the bladder. The other open-end of the bladder is sealed against two matching bevelled flanges—see Fig. 7.7b. The mating surfaces compress the silicone and seal the bladder. A similar arrangement is shown in Ref. [4]. At the knotted end, a blanking plate and an end cap with a hemispherical cavity are inserted in the mould to control the expansion (and prevent splitting) of the silicone bladder during the moulding process.

To create the tube, four layers of twill weave prepreg are again used—see Sect. 7.2. The prepreg plies are placed on top of each other with a slight offset (see Fig. 7.8a), *rolled* to remove trapped air and then loosely wrapped into a tubular preform around the bladder. The offset for each ply ensures a gradual transition between the prepreg

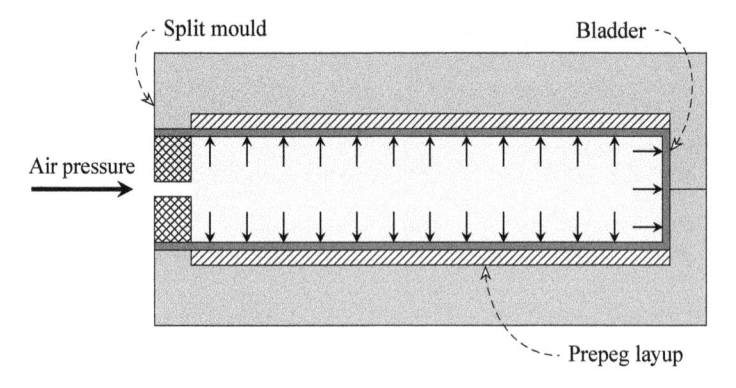

Fig. 7.6 Bladder moulding (schematic)

7.3 Bladder Moulding

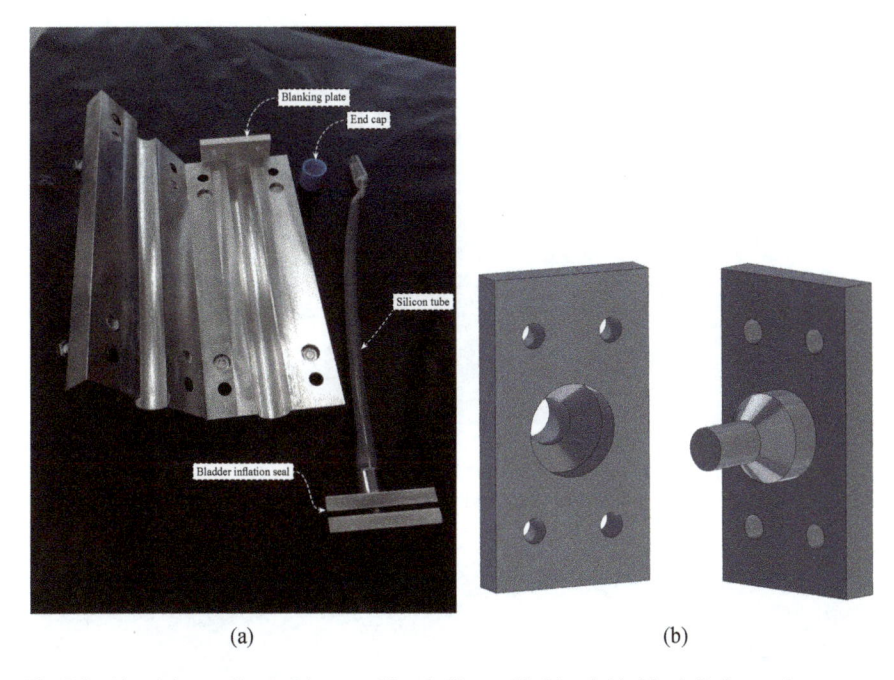

Fig. 7.7 Aluminium split mould: **a** mould and silicone bladder; **b** bladder inflation seal

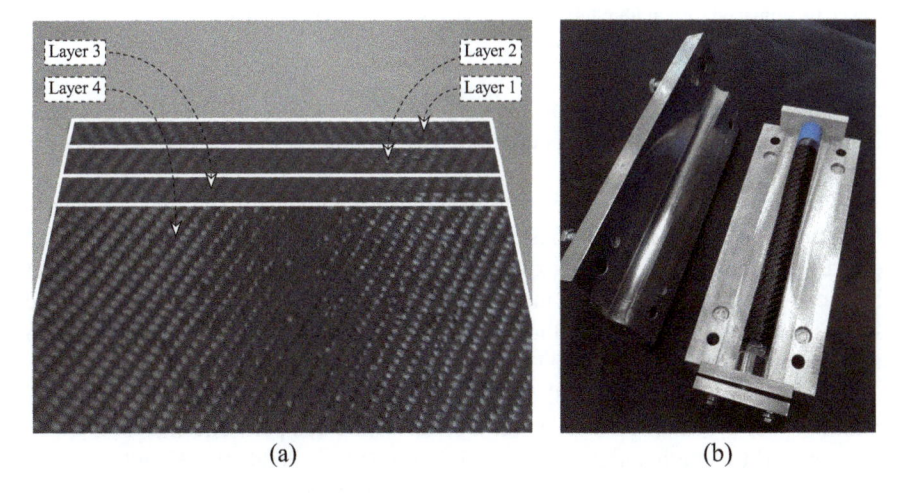

Fig. 7.8 Bladder Moulding: **a** prepreg layup; and **b** tube preform

layers. The prepreg layup and silicon bladder are then positioned in the mould as shown in Fig. 7.8b.

The mould is assembled and clamped shut. The inflation pressure is gradually increased up to 2 bar and the bladder is checked to ensure there are no leaks. The

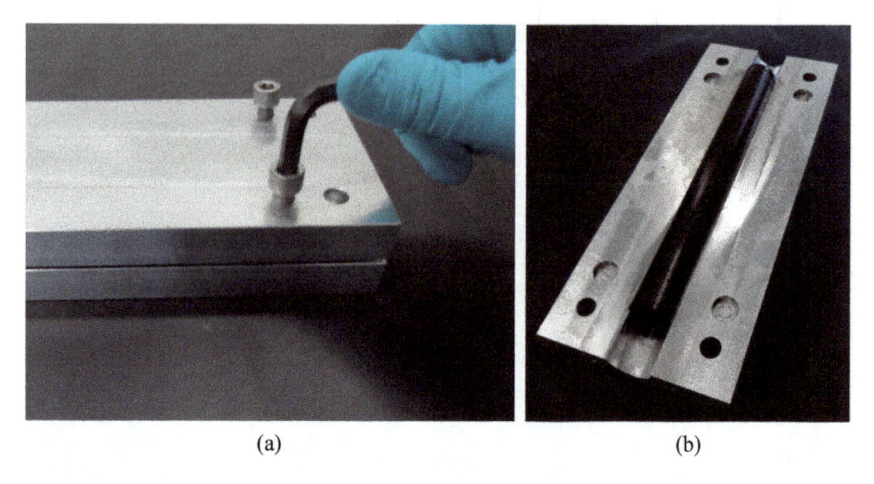

Fig. 7.9 Demoulding: **a** mould separation; and **b** cured composite tube

Fig. 7.10 Finished carbon fibre tube manufactured using bladder moulding

mould is placed in the oven and the prepreg is cured according to the supplier's instructions. Bladder pressure is continuously monitored to ensure no leaks occur during the curing process.

After curing, the carbon fibre-reinforced tube is demoulded. The silicone bladder is deflated, and the mould is deconstructed. The mould is separated here with minimal effort, but if some difficulty is experienced splitting the mould halves, four socket head cap screws can be inserted into the threaded holes in the upper mould half. The screws can be tightened against the recesses in the lower mould half and the stubborn mould can be jacked apart (Fig. 7.9a) to reveal the composite tube (see Fig. 7.9b).

The cured part has a smooth surface finish with few imperfections and thus, minimal post-processing is needed for a bladder moulded part. The ends of the tube are trimmed to length and the part is lightly sanded (to P800, or 400–500 grit). To finish the tube, the same clear coat used earlier in the mandrel lamination process is introduced to again provide a smooth surface over the carbon fibre aesthetic—see Fig. 7.10.

Fig. 7.11 Schematic of a helical spring (with dimensions)

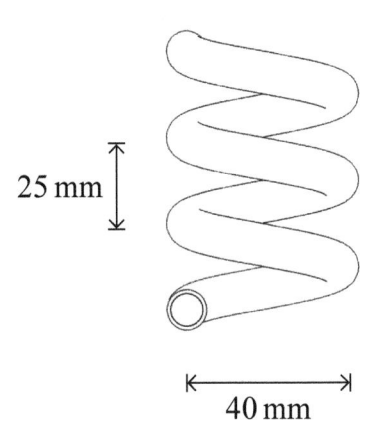

25 mm

40 mm

7.4 Practical Task: a Helical Spring

In this section, we extend the manufacture of hollow structures to consider a more complicated shape—a helical (coil) spring. To create a hollow tube, we use carbon fibre braided sleeves over a wax rod to create a *mouldless* composite construction. The spring coil preform is then achieved by spiralling the mouldless construction around a 3D printed mandrel.

Design Specification. The helical compression spring shown in Fig. 7.11 is to be manufactured using a carbon fibre-reinforced epoxy composite. The spring design requires a compression stiffness (referred to as the *spring rate*), k, of greater than 20 N/mm and a displacement or travel, x, of at least 5 mm. The spring should have at least 3 turns (windings), a mean coil diameter of 40 mm and a spring pitch of 25 mm.

Composite Design. The helical spring is fabricated using a wet layup process with carbon fibre (braided sleeve, 350 g/m² [22]) in an epoxy matrix [23]. Two plies of carbon fibre are used to fabricate a hollow circular cross-section—see Example 7.2.

Example 7.2

How many carbon fibre plies are needed to manufacture the helical spring in Fig. 7.11? The spring rate, k, must be greater than 20 N/mm and the deflection should be at least 5 mm.

For a helical spring with a hollow cross-section, assume the spring rate and deflection can be approximated by

$$k = \frac{G_c \times (d_o^4 - d_i^4)}{8 C^3 N}$$

and

$$x = \frac{\tau_c \pi C^2 N}{d_o G_c}$$

Note. G_c is the shear modulus of the composite and τ_c is the shear stress. The parameters d_o and d_i are the outer and inner diameters of the spring cross-section, respectively, C is the coil mean diameter and N is the number of turns.

Solution 7.2

In this example, we assume the helical spring has an inner diameter of 10 mm. Carbon fibre braided sleeves are used as the reinforcing fibres, but it is noted that other fibre selections could produce an equally successful outcome. The braids are selected in this instance to provide fibre angles of approximately $\pm 45°$ to the spiralling spring axis. In doing so, the fibres offer the maximum resistance to the torsion load applied to the spring's cross-section—see Sect. 2.6.

The fibre and matrix properties assumed here are shown in Table 7.1.

Table 7.1 Fibre and matrix properties

Material	Elastic modulus (GPa)	Tensile strength (MPa)	Poisson's ratio
Carbon fibre	240 [22]	3800 [22]	0.22 [24]
Epoxy	3.65 [23]	83.3 [23]	0.39 [24]

The range of fibre volume fractions for wet layup of woven fabrics (including braids) is 0.2–0.4—see Table 3.1. Therefore, we initially take a mid-range $V_f = 0.3$ for our design estimates.

Since the shear performance of a laminate with fibres orientated at +45. or -45. is the same, we approximate the woven laminate as a UD composite (with the correct Vf) offset at 45. to the shear loading.

Spring Rate. The spring rate is dependent upon the shear modulus of the FRC, G_{c45}, via

$$k = \frac{G_{c45} \times (d_o^4 - d_i^4)}{8 C^3 N}$$

7.4 Practical Task: a Helical Spring

The shear modulus (at $\pm 45°$), G_{c45}, is calculated from

$$\frac{1}{G_{c45°}} = \frac{1 + 2\nu_c}{E_{cl}} + \frac{1}{E_{ct}} \tag{7.3}$$

where the longitudinal modulus, E_{cl}, is

$$\begin{aligned}
E_{cl} &= E_m(1 - V_f) + E_f V_f \\
&= 3.65 \times 10^9 (1 - 0.3) + 240 \times 10^9 \times 0.3 \\
&= 74.6 \times 10^9 \text{ N/m}^2 = 74.6 \text{ GPa}
\end{aligned}$$

and the transverse modulus, E_{ct}, is

$$\begin{aligned}
\frac{1}{E_{ct}} &= \frac{(1 - V_f)}{E_m} + \frac{V_f}{E_f} \\
\frac{1}{E_{ct}} &= \frac{(1 - 0.3)}{3.65 \times 10^9} + \frac{0.3}{240 \times 10^9} \\
E_{ct} &= 5.2 \times 10^9 \text{ N/m}^2 = 5.2 \text{ GPa}
\end{aligned}$$

Poisson's ratio, ν_c, is given by

$$\begin{aligned}
\nu_c &= \nu_m(1 - V_f) + \nu_f V_f \\
&= 0.39 \times (1 - 0.3) + 0.20 \times 0.3 \\
&= 0.33
\end{aligned}$$

Thus,

$$\begin{aligned}
\frac{1}{G_{c\,45°}} &= \frac{1 + (2 \times 0.33)}{74.6 \times 10^9} + \frac{1}{5.2 \times 10^9} \\
G_{c\,45°} &= 4.7 \times 10^9 \text{ N/m}^2 = 4.7 \text{ GPa}
\end{aligned} \tag{7.4}$$

Taking $d_i = 10$ mm, we can rearrange the spring stiffness equation to calculate the minimum outer diameter, d_o, to be

$$\begin{aligned}
d_o &\geq \sqrt[4]{\frac{8 k C^3 n}{G_{c\,45°}} + d_i^4} \\
&\geq \sqrt[4]{\frac{8 \times 20000 \times 0.04^3 \times 3}{4.7 \times 10^9} + 0.01^4} \\
&\geq 0.0113 \text{ m} = 11.3 \text{ mm}
\end{aligned}$$

The thickness of each ply (sleeve) is

$$t = \frac{nA_w}{\rho_f V_f}$$
$$= \frac{1 \times 350}{1780 \times 0.3} = 0.66 \, \text{mm}$$

and hence, the number of sleeves is calculated to be

$$n = \frac{(11.3 - 10)/2}{0.66} = 0.98$$

Two carbon fibre sleeves are conservatively selected here. This is to account for a factor of safety (see Sect. 5.5) that addresses design and manufacturing inaccuracies and other uncertainties. The selection will yield an outer diameter of approximately 12.64 mm and a spring rate of

$$k = \frac{4.7 \times 10^9 \times (0.01264^4 - 0.01^4)}{8 \times 0.04^3 \times 3}$$
$$= 47509 \, \text{N/m} = 47.5 \, \text{N/mm} > 20 \, \text{N/mm} \checkmark$$

Deflection. The spring deflection at failure is dependent upon the shear modulus and shear strength of the braided composite, viz.

$$x = \frac{\tau^*_{c\,45°} \, \pi \, C^2 \, N}{d_o \, G_{c45°}}$$

As previously mentioned, the estimation of composite strength under *off-axis* loading is a complex task that is beyond the scope of this textbook—see Sect. 2.6. However, it has also been noted that the maximum shear strength occurs with fibre orientations at $\pm 45°$. Since the braided sleeves have fibres at $\pm 45°$, we can use the in-plane shear strength (τ^*_c) as a conservative estimate of $\tau^*_{c45°}$.

The in-plane shear strength is estimated from

$$\tau^*_c \approx \tau^*_m = 0.5 \times \sigma^*_m,$$
$$= 0.5 \times 83.3 \times 10^6$$
$$= 41.7 \times 10^6 = 41.7 \, \text{MPa}$$

Thus, at failure, the spring deflection is defined as

7.4 Practical Task: a Helical Spring

$$x > \frac{\tau_c^* \pi C^2 N}{d_o G_c 45°}$$

$$> \frac{(41.7 \times 10^6) \times \pi \times 0.04^2 \times 3}{0.01264 \times (4.7 \times 10^9)}$$

$$> 0.0106 \, \text{m} = 10.6 \, \text{mm}$$

This conservative (low bound) estimate of spring travel (using two braided sleeves) exceeds the minimum of 5 mm. ✓

Summary

The design estimates have indicated that two carbon fibre braided sleeves (A_w= 350 g/m²) will exceed the minimum stiffness of 20 N/mm, and permit more than 5 mm of spring travel (displacement). The estimates are based on a helical (coil) spring with a circular hollow cross-section and inner diameter of 10 mm.

Manufacturing Process. To make the mouldless construction for the helical spring, a wax rod of approximately 10 mm is used as the inner *hollow* core of the helical spring—see Fig. 7.12a. The carbon fibre braided sleeves (10 mm diameter, 350 g/m²) [22] are precut and the epoxy resin [23] is mixed in accordance with the supplier's instructions, assuming a wastage factor of 1.25—see Example 7.3. It should be noted that the braided sleeves are cut slightly longer than needed to allow the spring to be trimmed to length after the composite is demoulded.

Example 7.3

Consider Example 7.2. How much resin and how much hardener are needed to fabricate the helical spring (with two braided sleeves)?

Solution 7.3

The two carbon fibre braided sleeves (10 mm diameter) were cut to 700 mm long. The surface area of each sleeve was therefore $\pi \times 0.01 \, \text{m} \times 0.7 = 0.022 \, \text{m}^2$.

The fibre volume fraction was previously chosen based on the woven carbon fibre values for wet layup in Table 3.2, i.e. $V_f = 0.3$.

For two sleeves, $W_f = 0.38$ and hence

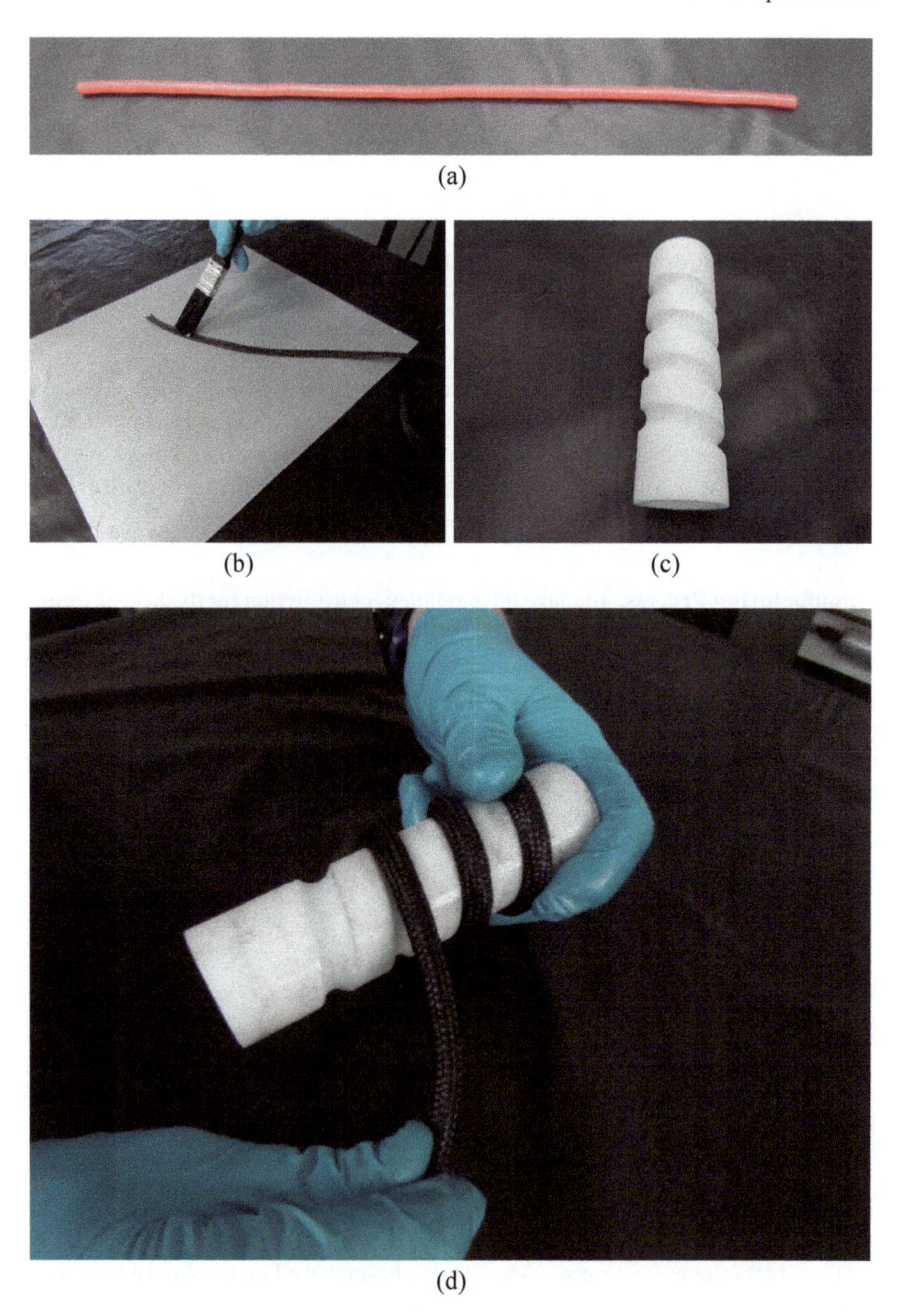

Fig. 7.12 Spring layup process: **a** wax rod; **b** *wet out* of braids; **c** 3D printed mandrel; and **d** creating the helical preform

7.4 Practical Task: a Helical Spring

$$\text{Resin and hardener mix (g)} = \frac{A \times n \times A_w \times (1 - W_f)}{W_f}$$

$$= \frac{0.022 \times 2 \times 350 \times (1 - 0.38)}{0.38} = 25.1\,\text{g}$$

To account for wastage, a factor of 1.25 (between 1 and 1.5) is used for the resin and hardener mix. Thus

$$\text{Resin and hardener mix (g)} = 75 \times 1.25 = 31.4\,\text{g}$$

The epoxy resin used in this case has a resin to hardener ratio of 4:1 [23], i.e. $4/5$ of resin and $1/5$ of hardener

$$\text{Resin (g)} = 31.4 \times 4/5 = 25.1\,\text{g}$$
$$\text{Hardener (g)} = 31.4 \times 1/5 = 6.3\,\text{g}$$

The carbon fibre braids are initially wet out (see Fig. 7.12b) before the sleeves are carefully slid over the top of the wax rod, one at a time. To assist with the layup, the sleeve diameter can be increased by compressing the tube's length and then, once in place, decreased by stretching the sleeve. Excess resin is removed before a spring preformed is created.

To make the spring preform, the carbon fibre sleeves (and internal wax rod) are gently formed around a 3D printed mandrel which has been pre-released with a wax release agent. The mandrel has a semi-circular groove that spirals around its circumference to define the spring shape—see Fig. 7.12c. The spiral preforming process is illustrated in Fig. 7.12d. The spring is left to cure on the mandrel for at least 24 h before demoulding. During demoulding, the spring is simply rotated (unscrewed) from the mandrel.

A post-cure temperature of 80 °C is used to melt the wax rod and hence, produce the hollow coil. The helical spring is then trimmed to length. Here, a spring that is slightly longer than three turns is manufactured—see Fig. 7.13. The excess length is used as a means to locate the spring in a prefabricated *housing* for physical testing.

Physical Testing. As mentioned, the FRC spring is located in a *spring housing* for testing on an Instron Universal Testing System (UTS)—see Fig. 7.14a. The 3D printed housing is used to constrain the lateral movement and rotation of the spring; motion is only unrestrained along the spring's compression axis. After a small preload is applied to the spring, the load-deflection characteristics are recorded at a displacement rate of 2 mm/min (up to 5.5 mm). The experimental measurements are compared to the spring rate and deflection requirements (from the design specification) in Fig. 7.14b. In the figure, the spring's performance is shown to exceed the design requirements.

Fig. 7.13 Helical spring after demoulding and trimming

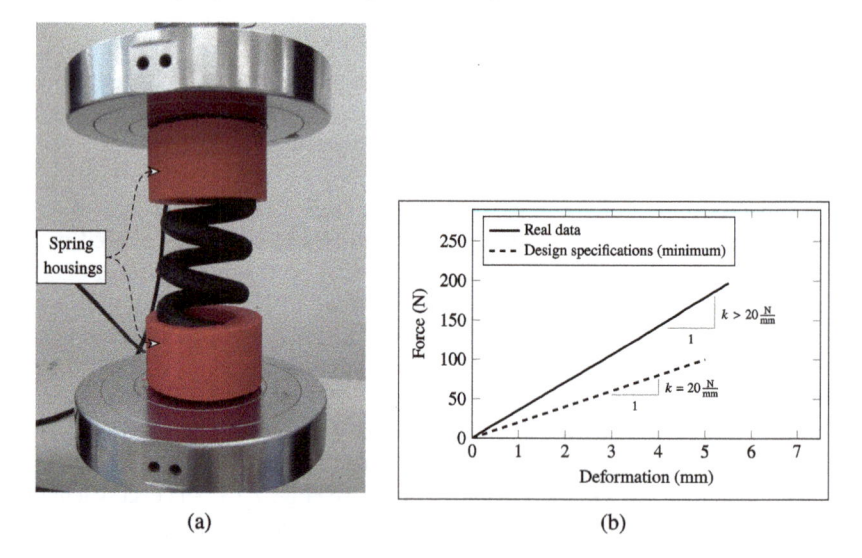

(a)   (b)

Fig. 7.14 Spring testing: **a** spring with the housing under compression, and **b** the mechanical response of the spring

7.5 Extension Task: a Bicycle Handlebar

Design Specification. As a further example, a carbon fibre-reinforced mountain bike (MTB) handlebar is fabricated (see Fig. 7.15) to satisfy the standard lateral bending test described in BS EN ISO 4210-5:2014[1] [25]. In the lateral bending test for an

---

[1] The lateral bending test is one of several tests specified in BS EN ISO 4210-5:2014 [25]. To be compliant, a bicycle handlebar must satisfy *all* relevant test procedures listed in the safety standard.

7.5 Extension Task: a Bicycle Handlebar

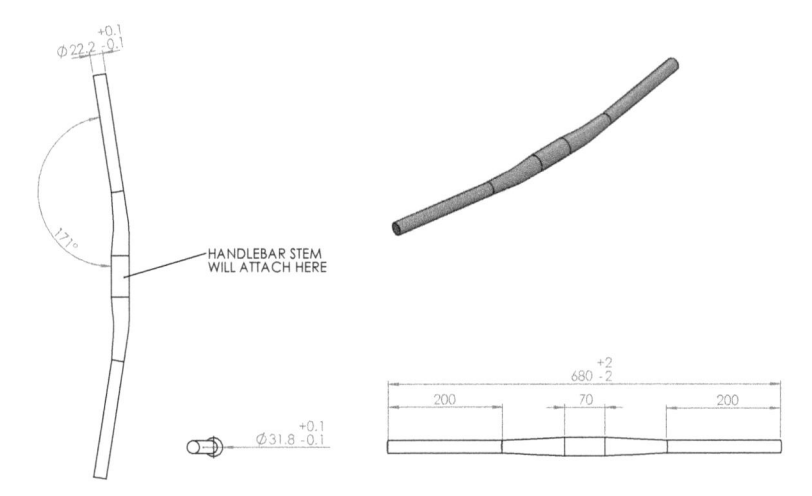

Fig. 7.15 MTB handlebar (all dimensions in millimetres). Reproduced from [26] with permission from N. Emerson

MTB, a force of 1000 N must be applied (for 1 min) at a distance of 50 mm from the free end of the handlebar. In a successful test, the FRC handlebars must not fail (i.e. either crack or fracture) under the loading conditions.

Composite Design. To create the MTB bicycle handlebar, six unidirectional plies and four plies of twill weave carbon fibre prepreg (both 200 g/m^2) are chosen. This selection is made based on design calculations and some initial assumptions. The result is far from an optimised solution but offers a starting point—see Example 7.4.

Here, we make and evaluate a handlebar prototype using carbon fibres orientated along the handlebar length and around the circumference *only*. Despite the obvious benefit for shear (transverse shear and torsion) performance, no ±45° fibres are used in this example. The inclusion of off-axis fibres is essential to improve handlebar performance and is left as an optimisation exercise for the reader. A more comprehensive analysis is possible with classical laminate theory [27], extended layer-wise theory [28] or finite element analysis (FEA) [29–31], but these more complex methods are beyond the scope of this textbook.

> **Example 7.4**
>
> A lateral bending test for the MTB bicycle handlebar (see Fig. 7.15) is schematically shown in Fig. 7.16. Carbon fibre prepreg will be used to manufacture the MTB handlebar. A load of 1 kN is applied at a distance of 50 mm from the free end of the handlebar (with a 28 mm eccentricity). What ply selection should be used? In this example, consideration should only be given to fibres orientated along the handlebar length and/or circumferentially around the tube.

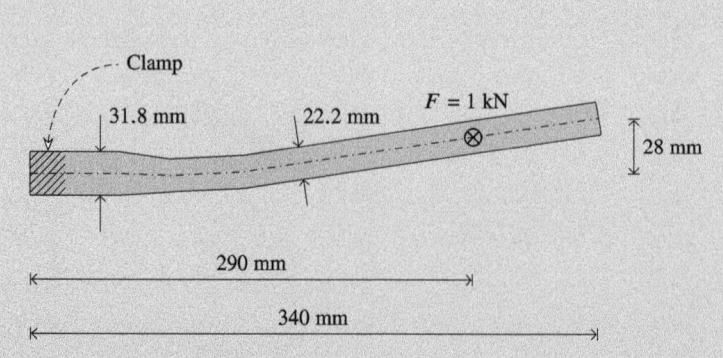

Fig. 7.16 Schematic of lateral bending test of MTB handlebar (top view) where a downward force F is being applied

Solution 7.4

Assumptions. Herein, the UD prepreg used in Chap. 5 is considered and hence, the mechanical properties in Ref. [32] are used for the design estimates—see Table 5.2. To account for design and manufacturing inaccuracies and other uncertainties, a factor of safety (FoS) of 2.0 is selected as a safety net.

The mechanical properties are:

$$E_{cl} = 131\,\text{GPa}$$
$$\sigma_{cl}^* = 2575\,\text{MPa}$$
$$\sigma_{cl,\,comp}^* = 1235\,\text{MPa}$$
$$G = 3.9\,\text{GPa}$$
$$\tau_c^* = 85.7\,\text{MPa}$$

Ply thickness, $t \approx 0.21$ mm   (...based on measurements in Table 5.1)

Structural Analysis. The 1 kN load causes bending, transverse shear and torsion of the MTB handlebar. The bending moment, shear force and torsion moment diagrams are shown in Fig. 7.17. In the same figure, the normal stress (from bending) and shear stresses (from transverse shear force and torsional) are also illustrated.

Normal (Bending) Stresses. The bending moment, M, increases linearly from the load application point to the clamp (constrained face). Along the length of the handlebar, the cross-sectional properties change. The diameter at the free end is 22.2 mm, whilst the diameter is 31.8 mm at the stem clamp. Thus, as a first approximation, it is conservative to assume that the maximum bending moment (at the stem clamp) acts on the minimum tube diameter (i.e. 22.2 mm).

7.5 Extension Task: a Bicycle Handlebar

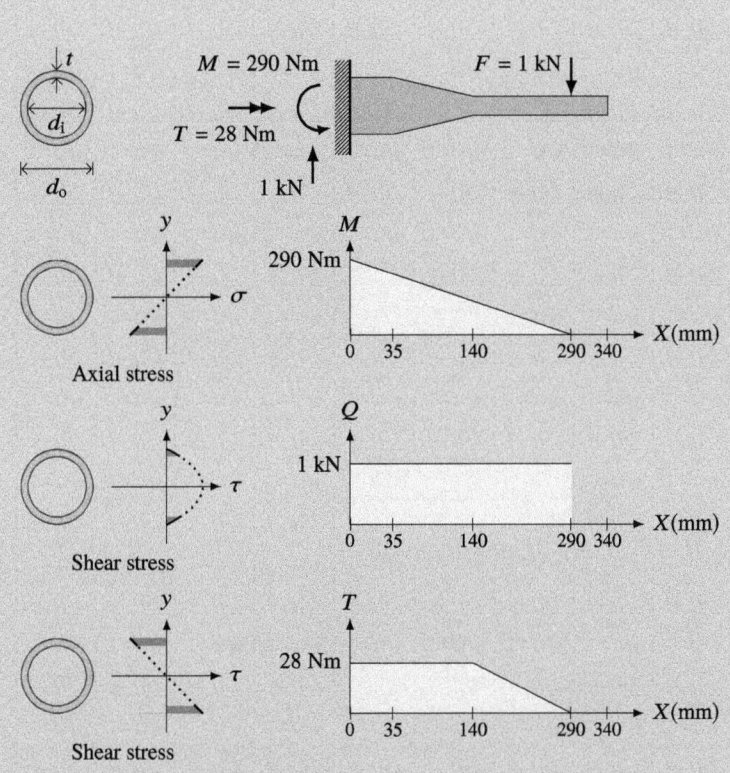

Fig. 7.17 Simplified model of the handlebar and the respective shear force (Q), bending moment (M) and torsion (T) diagrams

Assuming a unidirectional composite:

$$\sigma = \frac{My}{I}$$

$$\therefore \frac{\sigma_{cl}^*}{FoS} > \frac{My}{I} = \frac{My}{\frac{\pi(d_o^4 - d_i^4)}{64}}$$

$$d_i < \sqrt[4]{d_o^4 - \frac{64My \times FoS}{\pi \sigma_{cl,comp}^*}} \quad \text{(assuming } \sigma_{cl,comp}^* \text{ as the lower value)}$$

$$< \sqrt[4]{(22.2 \times 10^{-3})^4 - \frac{64 \times 290 \times 11.1 \times 10^{-3} \times 2.0}{\pi \times 1235 \times 10^6}} = 0.01923 \text{ m}$$

$$t = \frac{0.0222 - 0.01923}{2} = 0.00149 \text{ m} = 1.49 \text{ mm}$$

$$\therefore \text{No. of UD plies} = \frac{1.49}{0.21} = 7.1 \text{ plies (assuming} \approx 0.21 \text{ mm/ply)}$$

Thus, eight UD plies (*or the equivalent*) are needed to resist the bending moment.

Transverse Shear Stresses. The shear force, Q, is constant (1 kN) along the length of the handlebar so we can assume this load acts on the smallest section (22.2 mm). Note that τ_c^* is assumed to be 85.7 MPa.

For a unidirectional composite

$$\tau_{avg} = \frac{Q}{A}$$

$$\tau_{max} \approx \frac{2Q}{A} \text{ (assuming the tube is thin} - \text{walled)}$$

$$\therefore \frac{\tau_c^*}{\text{FoS}} > \frac{2Q}{A} = \frac{2Q}{\frac{\pi(d_o^2 - d_i^2)}{4}}$$

$$d_i < \sqrt{d_o^2 - \frac{8Q \times \text{FoS}}{\pi \tau_c^*}} = 0.02082 \text{ m} = 20.82 \text{ mm}$$

$$t = \frac{0.0222 - 0.02082}{2} = 0.00069 \text{ m} = 0.69 \text{ mm}$$

$$\therefore \text{ No. of UD plies} = \frac{0.69}{0.21} \approx 3.3 \text{ plies}$$

Note. To simplify the calculations, the MTB handlebars are considered to be analogous to a *thin-walled* structure. This assumption is usually only applied when the inner tube radius is at least 10 times larger than the thickness ($r/t > 10$ [33]).

Torsional Stresses. The torsional moment rises to a maximum (28 N·m) at the start of the tapered section and then remains at this maximum up to the stem clamp. Again, we assume the maximum load acts on the minimum section.

Thus, for a unidirectional composite

$$\tau = \frac{Tr}{J}$$

$$\therefore \frac{\tau_c^*}{\text{FoS}} > \frac{Tr}{J} = \frac{Tr}{\frac{\pi(d_o^4 - d_i^4)}{32}}$$

$$d_i < \sqrt[4]{d_o^4 - \frac{32Tr \times \text{FoS}}{\pi \tau_c^*}}$$

$$d_i < \sqrt[4]{(22.2 \times 10^{-3})^4 - \frac{32 \times 28 \times 11.1 \times 10^{-3} \times 2.0}{\pi \times 85.7 \times 10^6}}$$

$$d_i < 0.02028 \text{ m} = 20.28 \text{ mm}$$

7.5 Extension Task: a Bicycle Handlebar

$$t = \frac{0.0222 - 0.02028}{2} = 0.00096 \, \text{m} = 0.96 \, \text{mm}$$

$$\therefore \text{No. of UD plies} = \frac{0.96}{0.21} \approx 4.6 \, \text{plies}$$

Now, considering the *Principle of Superposition* for both sources of shear stresses, it follows that

No. of UD plies = 3.3 (transverse shear) + 4.6 (torsion) = 7.9

Thus, eight UD plies (*or the equivalent*) are needed to resist the shear stresses.

Normal (Hoop) Stresses. In addition to the normal (bending) and shear stresses mentioned above, consideration must be given to the stress concentrations that arise at the stem clamp and directly under the applied 1 kN load. Both of these are likely to cause normal stresses in the hoop (transverse) direction. These stresses could cause localised crushing issues. To minimise these local effects, we must introduce extra plies with fibres orientated circumferentially around the tube. Here, we assume that the equivalent of two UD plies will be sufficient.

So, summarising the ply selection... To resist the bending moment, we require at least eight UD plies, whilst shear (due to torsion and transverse shear) also requires the equivalent of eight UD plies. To account for localised crushing, we assume the equivalent of two circumferentially orientated UD plies is needed.

Failure analysis. Laminate failure is assumed to occur when a stress, either parallel or perpendicular to the fibre orientations (namely, σ_{cl}, σ_{ct} or τ_c), exceeds the critical (strength) value; this is the basis of the maximum stress criterion [24]. The normal and shear stresses are treated independently from one another, viz.:

$$\sigma_{cl} < \sigma_{cl}^*.$$
$$\sigma_{ct} < \sigma_{ct}^*.$$
$$\tau_c < \tau_c^*.$$

Here, to satisfy the maximum stress criteria, we select:

- *six UD plies* with the fibres oriented along the handlebar length.
- *four woven plies* with warp fibres along the length and weft circumferential.

In making this layup selection, some *judgement* has been made with regard to the equivalence of UD and woven plies in the context of the handlebar stresses. A woven ply is assumed to offer about half the bending performance of a UD ply along the length of the handlebar, as well as half the UD stiffness and strength in the hoop orientation (neglecting the reduction in V_f for woven prepreg). Moreover, the woven plies are assumed to provide the equivalent shear performance of a UD layer [34]—see Sect. 2.7 for further information.

Summary. Based on our design estimates, the handlebar layup (i.e. 6 UD and 4 woven prepreg plies) should satisfy the requisite lateral bending load case, offering the equivalent of

- *eight* UD plies to resist the bending moment, meeting the requisite eight plies: 6 UD +(4 woven × 0.5) = 8 ✓
- *Ten* UD plies to resist the shear stresses, exceeding the requisite eight plies: 6 UD + (4 woven × 1) = 10 ✓
- *Two* UD plies to resist localised crushing: (4 woven × 0.5) = 2 ✓

Note. This is *only* an initial prototyping exercise and therefore many approximations (and judgements) have been made to simplify the above calculations.

Manufacturing Process. The composite handlebar is manufactured using the bladder moulding method described in Sect. 7.3. The mould used is a two-part split aluminium mould (6061) with a similar bladder inflation seal, blanking plate and end cap arrangement—see Fig. 7.18. The split mould is manufactured using specialist CNC machine tools. Prior to prepreg layup, the mould is released with six coats of high-temperature wax suitable for temperatures up to 120 °C.

Here, the handlebar is considered as three sections: the central (stem) clamping section; and the two handle sections, including the tapered region connecting the handles to the stem section—see Fig. 7.19. To minimise the effects of sectioning, the prepreg is cut oversize; the prepreg is cut to be longer and wider than needed for the handlebar. An overlap of at least 0.5 inches (≈12.5 mm) [35] is recommended in a *sectioned* layup process to maintain the strength of the composite. Here, the over-

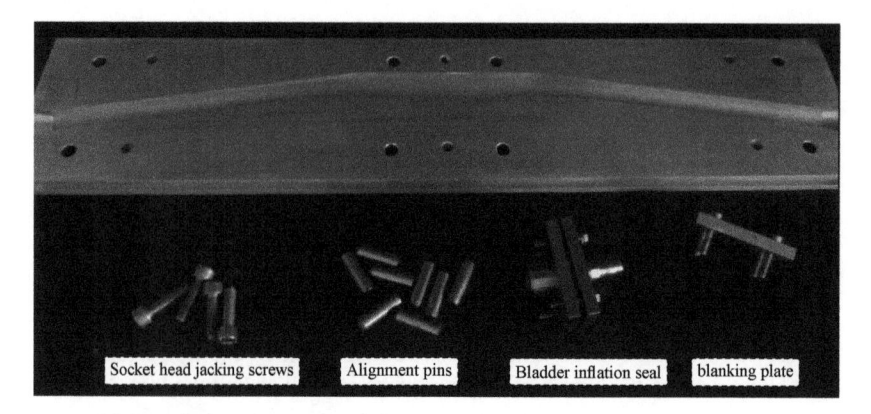

Fig. 7.18 Handlebar 'split' mould (only half is shown) and ancillaries

7.5 Extension Task: a Bicycle Handlebar

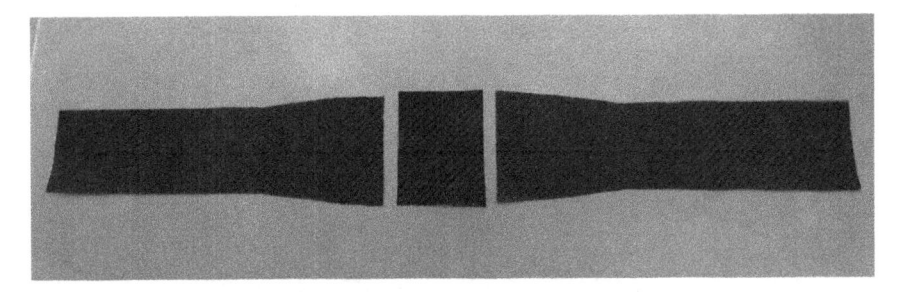

Fig. 7.19 Prepreg sectioning of handlebar (no dimensions)

laps are cut slightly larger than these minimum requirements, adding weight to the handlebar but offering confidence in the sectioning method used during fabrication.

The plies for each of the prepreg sections are initially consolidated on a flat glass plate with a roller (see Fig. 7.20a) and then preformed into a tube—see Fig. 7.20b. The inflation bladder is threaded through each of the tube preforms. The preforms are positioned in one half of the handlebar mould with the wider end of the handle sections inserted into the stem section as shown in Fig. 7.20c. The mould is closed and the locating pins are inserted to align the mould halves. The inflation assembly and blanking plate are then installed as shown in Fig. 7.20d. Finally, the two mould halves are clamped shut.

The inflation bladder is gradually pressurised to consolidate the prepreg against the mould cavity. The bladder is checked to ensure no leaks are present before commencing the curing process.

The mould is placed in an oven, and the prepreg is cured and post-cured in line with the supplier's instructions. The inflation pressure is continuously monitored to ensure no leaks occur during the curing process.

During the demoulding process, the silicone bladder is deflated and the mould halves are separated. If the split mould is difficult to part, the cap head bolts can be inserted into the threaded holes on one of the mould halves and tightened against the recess of the other mould half to jack the mould apart—as previously discussed; see Fig. 7.9a. The composite handlebar is then easily removed from the mould as shown in Fig. 7.21.

The handlebar is cut to length and weighed before physical testing commences; no surface finish (paint or clear coat) is applied in this instance. The mass of the handlebar (without optimisation) is measured to be approximately 230 g. This is comparable to other carbon fibre handlebars currently on the market [36, 37], and is notably lighter than aluminium handlebars which often weigh in excess of 300 g [38, 39].

Physical Testing. The handlebar is clamped in a stem extension on a lateral bending test rig. The test rig is bolted to an Instron UTS as shown in Fig. 7.22a. The handlebar is aligned perpendicular to the stem axis as described in the standard [25], and a monotonic load is applied (up to 1000 N) at a distance of 50 mm from the free end. Load-deflection measurements are recorded at a displacement rate of 5 $^{mm}/_{min}$.

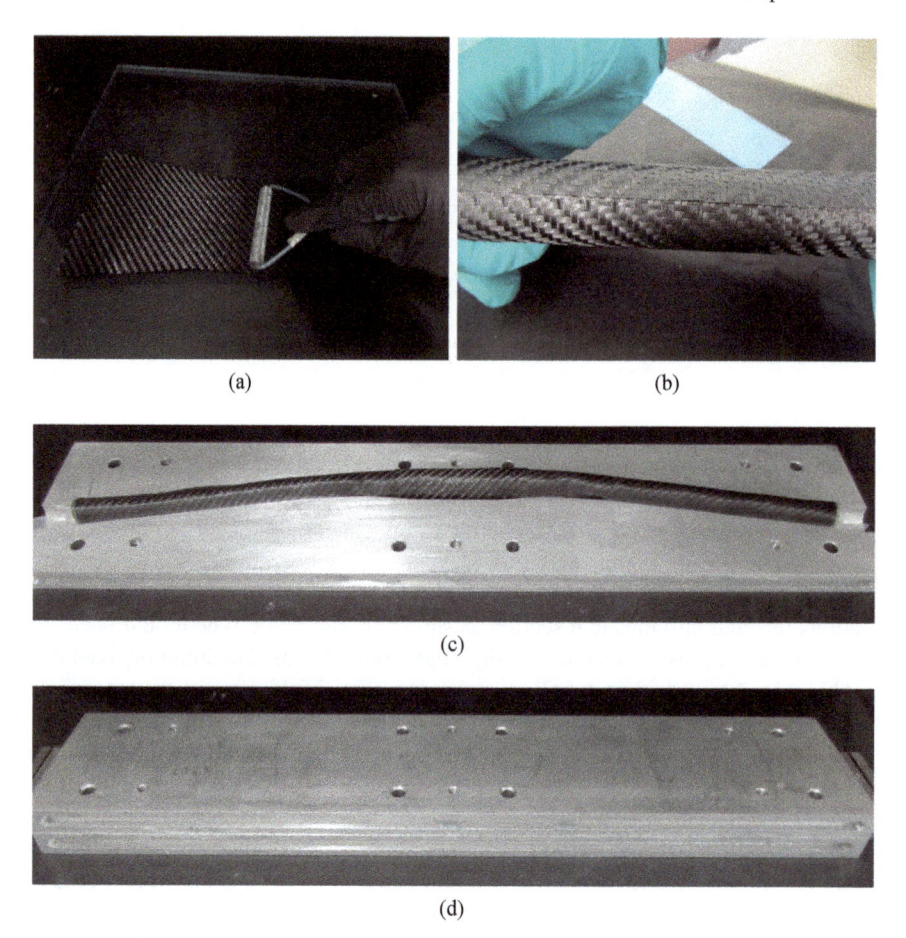

Fig. 7.20 Handlebar layup process: **a** initial ply consolidation; **b** a tube preform; **c** preform layup; and **d** mould assembled

Fig. 7.21 Demoulded handlebar

7.5 Extension Task: a Bicycle Handlebar

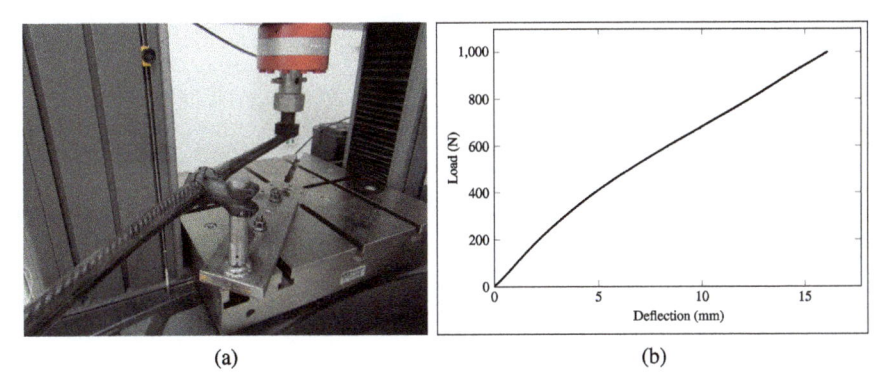

Fig. 7.22 Lateral bending test: **a** testing setup; **b** Load-deflection characteristics of composite handlebar

The experimental data is shown in Fig. 7.22b. The maximum load is maintained for at least 1 min without the handlebar cracking or fracturing. This is a successful outcome for a first structural prototype, but a lighter handlebar is possible if $\pm 45°$ fibres are used.

7.6 Summary

Hollow sections minimise the volume of material, and hence can be used to make lightweight and economical structures.

Two methods for creating hollow FRC tubes are:

- Mandrel lamination.
- Bladder moulding.

Mandrel lamination usually has lower tooling (mandrel) costs, but it tends to produce a hollow form with a less than ideal visual appearance on the external surface. In contrast, the bladder moulding process produces tight (external) tolerances and a near-perfect visual appearance, but usually at the expense of higher mould tool costs.

7.7 Questions

Questions 7.1 Define *mandrel lamination* and *bladder moulding*.

Questions 7.2 Describe three methods that could be used to assist mandrel removal in open-ended parts.

Questions 7.3 Why is mandrel lamination more complicated for closed-ended structures than for their open-ended counterparts?

Questions 7.4 Describe two methods that could be used to remove a mandrel from a closed-ended part.

Questions 7.5 What is *shrink tube* and how does it work? How does it differ from *shrink tape*?

Questions 7.6 In bladder moulding, why are silicone and latex excellent choices for a bladder?

Questions 7.7 In a split mould, what is the purpose of the alignment pins? Why is it still important to clamp the mould halves shut?

7.8 Problems

Problem 7.1 A unidirectional glass fibre-reinforced square hollow section (SHS) is 40 mm × 40 mm × 1.2 mm. The glass fibres have an elastic modulus of 69 GPa and a tensile strength of 2500 MPa; the polyester matrix has a modulus of 2.5 GPa and a strength of 50 MPa. If the SHS is subjected to a tensile load of 10 kN, calculate

a. The axial stress in the SHS.
b. The Factor of Safety (FoS), assuming a fibre volume fraction of 0.35.

Answer. a. 53.7 MPa; and b. 16.3.

Problem 7.2 A unidirectional hollow circular tube is clamped at one end (i.e. a cantilever). If the outer diameter of the tube is 30 mm and the wall thickness is 1 mm, calculate the stress in the *thin-walled* tube when a torque of 10 N·m is applied at the free end. If the in-plane shear strength of the composite is 40 MPa, will the composite fail?

Note. The polar second moment of area for a tube is $J = \frac{\pi}{32}(d_o^4 - d_i^4)$, where d_o is the outer diameter and d_i is the inner diameter of the tube.

Answer. 7.8 MPa. No ($\tau_c < \tau_c^*$).

Problem 7.3 A tubular shaft is to be designed with an outside diameter of 25 mm and a length of 0.5 m. The tube is manufactured from a unidirectional fibre-epoxy prepreg. The tensile modulus of the carbon-epoxy prepreg is 110 GPa, and the tensile strength and in-plane shear strength are 1800 MPa and 60 MPa, respectively. In service, the tubular shaft is subjected to a three-point bending test with a midspan load of 1kN. Determine the minimum number of 0.21 mm plies that are needed, to ensure

7.8 Problems

a. Longitudinal (tensile or compressive failure) does not occur.
b. Shear failure of the *thin-walled* tube does not occur.
c. The tubular shaft does not exhibit a midspan deflection of more than 10 mm.

Note. The second moment of area for a shaft is $I = \pi \frac{d_o^4 - d_i^4}{64}$, where d_i and d_o are, respectively, the inner and outer diameters; maximum deflection at the midspan of the three-point bending test is $\delta_{max} = \frac{Fl^3}{48EI}$, where l is the length of the beam, and F is the point load.

Answer. a. 4 plies; b. 2 plies; and c. 7 plies.

References

1. Strong AB (2008) Fundamentals of composites manufacturing: materials, methods and applications, 2nd edn. Society of Manufacturing Engineers, Dearborn, Mich
2. Hibbeler RC (2014) Statics and mechanics of materials, 4th edn. Pearson, Upper Saddle River
3. Wanberg J (2009) Composite materials: fabrication handbook #1, composite garage series, vol 1. Wolfgang Publications, Stillwater, Minnesota
4. Wanberg J (2010) Composite materials: fabrication handbook #2. Composite garage series. Wolfgang Publications, Stillwater, Minnesota
5. Astrom BT (2018) Manufacturing of polymer composites, 2nd edn. Routledge, Boca Raton
6. Hollaway L (1994) Handbook of polymer composites for engineers. Woodhead Publishing Ltd, Cambridge
7. Lee SM (1992) Handbook of composite reinforcements. VCH, New York
8. Campbell FC (2004) Manfacturing processes for advanced composites. Elsevier Advanced Technology, Oxford
9. Akovali G (2001) Handbook of composite fabrication. Woodhead Publishing, Shawbury
10. Aird F (2014) Fiberglass and other composite materialshp1498: a guide to high performance non-metallic materials for automotiveracing and marine use includes fiberglass, kevlar, carbon fiber, molds, structures and materials. HP Books, New York
11. Barbero EJ (2017) Introduction to composite materials design, composite materials, 3rd edn. CRC Press, Boca Raton
12. Silcock MD, Garschke C, Hall W, Fox BL (2007) Rapid composite tube manufacture utilizing the quicksteptm process. J Compos Mater 41(8):965–978. https://doi.org/10.1177/0021998306067261
13. Leslie JC (1970) Precision molding of advanced fiber structures. In: Volume 1B: general. American Society of Mechanical Engineers. https://doi.org/10.1115/70-GT-126
14. Francucci G, Palmer S, Hall W (2018) External compaction pressure over vacuum-bagged composite parts: effect on the quality of flax fiber/epoxy laminates. J Compos Mater 52(1):3–15. https://doi.org/10.1177/0021998317701998
15. Mazumdar SK (2002) Composites manufacturing: materials, product, and process engineering/Sanjay K. CRC Press, Boca Raton
16. Shenoi RA, Wellicome JF (1993) Composite materials in maritime structures, Cambridge ocean technology series, vols 4,5. CUP, Cambridge
17. Fiore V, Valenza A (2013) Epoxy resins as a matrix material in advanced fiber-reinforced polymer (frp) composites. In: Advanced fibre-reinforced polymer (FRP) composites for structural applications, Elsevier, pp 88–121. https://doi.org/10.1533/9780857098641.1.88
18. Ahmed A, Tavakol B, Das R, Joven R, Roozbehjavan P, Minaie B (2012) Study of thermal expansion in carbon fiber-reinforced polymer composites. In: SAMPE international symposium proceedings

19. Summerscales J, Cullen R (02/10/2019) Private correspondence
20. Callister WD, Rethwisch DG (2018) Materials science and engineering: an introduction, 10th edn. Wiley, Hoboken
21. Anderson JP, Altan MC (2014) Bladder assisted composite manufacturing (bacm): challenges and opportunities. In: Polymer processing society Europe-Africa conference. https://doi.org/10.13140/2.1.2139.6169
22. Dowaksa (2016) 3k a-38: Technical data sheet. https://www.carbononline.nl/wp-content/uploads/2020/05/Dowasksa-3K-A-38-7.pdf
23. Kinetix (2018) Engineering data: R240 wet-preg. http://atlcomposites.com.au/icart/products/7/images/main/KINETIX240Pre-PregEngineeringData.pdf
24. Hull D, Clyne TW (1996) An introduction to composite materials, 2nd edn. Cambridge solid state science series. Cambridge University Press, Cambridge
25. British Standards (2014) Cycles - safety requirements for bicycles
26. Emerson N (08/06/2018) Private correspondence
27. Altenbach H, Altenbach JW, Kissing W (2004) Mechanics of composite structural elements. Foundations of engineering mechanics. Springer, Berlin
28. Javanbakht Z, Aßmus M, Naumenko K, Öchsner A, Altenbach H (2019) On thermal strains and residual stresses in the linear theory of anti–sandwiches. ZAMM - J Appl Math Mech/Zeitschrift für Angewandte Mathematik und Mechanik 99(8). https://doi.org/10.1002/ZAMM.201900062
29. Javanbakht Z, Öchsner A (2017) Advanced finite element simulation with MSC Marc: application of user subroutines. Springer, Cham
30. Javanbakht Z, Öchsner A (2018) Computational statics revision course. Springer International Publishing, Cham. https://doi.org/10.1007/978-3-319-67462-9
31. Javanbakht Z, Hall W, Öchsner A (2017) Computational evaluation of transverse thermal conductivity of natural fiber composites. In: Öchsner A, Altenbach H (eds) Improved performance of materials: design and experimental approaches. Advanced structured materials, vol 72, Springer, Cham, pp 197–206. https://doi.org/10.1007/978-3-319-59590-0
32. Solvay (2021) Technical data sheet: Vtm(mohana)260 series (prepeg). https://catalogservice.solvay.com/
33. Govindjee S (2012) Engineering mechanics of deformable solids: a presentation with exercises. OUP Oxford, Oxford
34. Medina C, Canales C, Arango C, Flores P (2014) The influence of carbon fabric weave on the in-plane shear mechanical performance of epoxy fiber-reinforced laminates. J Compos Mater 48(23):2871–2878. https://doi.org/10.1177/0021998313503026
35. Jin H, Nelson K, Werner BT, Briggs T (2018) Mechanical strength of composites with different overlap lengths: (No. SAND2018-10594). Sandia National Lab (SNL-CA), Livermore, CA (United States). https://doi.org/10.2172/1488647
36. Hope Technology (2020) Carbon handlebar 31.8mm. https://www.hopetech.com/products/controls/handlebars/carbon-handlebar-318mm/
37. Renthal (2020) Fatbar carbon 10mm rise. https://www.renthal.com/cycle/handlebars/fatbar-carbon/fbc-10mm
38. Deity Components (2020) Deity blacklabel 800 handlebar: 15mm rise. https://www.deitycomponents.com/store/p8/DEITY_BLACKLABEL_800_HANDLEBAR_15mm_RISE.html
39. Renthal (2020) Fatbar 10mm rise. https://www.renthal.com/cycle/fatbar/fatbar-10

Index

A
Additive manufacturing, 86
Air-assisted demoulding, 109
Anisotropy, 13, 14
Axial, *see* Longitudinal

B
Bladder, 112
Bladder moulding, 105, 112, 128
Bleeder, 56
Bleeder/breather fabric, 59, 63
Breach unit, 59
Breakdown method, 98, 106
Breather, 56
Bridging, 84

C
Catalyst, 41, 44
Caul plate, 37
Classic laminate theory, 123
Coefficient of Thermal Expansion (CTE), 109
Composite
    consolidation, 56
    design, 13
    failure mode, 21
    machining, 47
    shear failure, 21
    shear modulus, 18, 20
    shear strength, 23
    strength, 21, 23
    tooling, 89
    unidirectional, 14, 16, 21, 25, 35
Consolidation pressure, 35
Coupling stress, 6
Crosslinking, 42, 43
Curing, 42, 107, 114

D
Debulking, 56
Demoulding, 42, 46, 108, 109, 112, 129
Dissolvable mandrel, *see* Mouldless construction
Draft angle, 82, 85
Drapeabilty, 6
Dry fibre weight, 26
Dwell time, 63

E
Epoxy, 2, 35, 42, 45, 56
Exothermic reaction, 42

F
Fabric, 5
    braided, 7
    plain, 6
    random mat, 5, 7, 13
    satin, 6
    twill, 6
    woven, 6, 25
Factor of safety, 77
Failure strain, 15
Fibre

aramid, 5
bundle, 5
carbon, 4, 35, 56
glass, 4
microbuckling, 22
orientation, 7, 20, 35, 55
random, 7, 25, 97
unidirectional, 5
volume fraction, 16, 35, 72
Finite element analysis, 123
Free radical, 44
Freezer-life, 62

G
Gelcoat, 9, 39, 48
Green stage, 42

H
Hand lamination, *see* Wet layup
Hardener, 41, 46
Hard tooling, 89
Heat shrink tube, 107
Hollow section, 105

I
Initiator, 44
Interface failure, 23
Inverse rule of mixture, 17, 76

K
Kelly-Tyson model, 21, 22, 76

L
Laminate, 7, 13, 16
Layup moulding, *see* Wet layup
Leak rate, 60
Longitudinal, 14
 axis, 7
 elastic modulus, 14, 20
 failure, 21, 22
 load, 24
 modulus, 72
 strength, 14, 22, 72

M
Mandrel, 106, 108
Mandrel lamination, 105
Master, *see* Plug

Modelling clay, 92
Mould, 37, 82
 female, 37, 81, 85
 matched-die, 37
Mouldability, 81
Mouldless construction, 106

O
Orientation efficiency, 25
Out-time, 62

P
Packing arrangement, 28
Parting agent, *see* Release agent
Pattern, *see* Plug
Pattern removal, 89
Pleat, 61
Plug, 85
 flexibility, 88, 91
Poisson's ratio, 20
Polyester, 2, 43
Polymer
 thermoplastic, 2, 86
 thermoset, 2
Polymerisation, 43
Polyvinyl Alcohol (PVA), 38
Post-curing, 42
Prepreg, 62, 105, 107, 112
Prepreg layup, 34
Prepreg moulding, *see* Prepreg layup

Q
Quasi-isotropy, 13

R
Reactive site, 45
Release agent, 38
Resin wastage, 41
Resin weight fraction, 26
Reuss model, *see* Inverse rule of mixture
Risk, 33
Rule of mixture, 15, 76

S
Sealant, 56, 61
Seasoned mould, 38
Shrink tape, 108
Styrene, 43
Surface finish, 37, 48, 85, 88, 89, 111

Surface texture, *see* Surface finish
Swiss roll layup, 107

T
Tensile test, 72
Tooling, 37
Transverse, 14
   elastic modulus, 17, 18, 20
   failure, 21, 23
   load, 24
   strength, 72
Tresca critertion, 23

U
Unidirectional laminates, 72

Unsaturated polymer, 43

V
Vacuum bagging, 34, 56, 61
Vacuum connector, *see* Breach unit
Vacuum valve, *see* Breach unit
Vinylester, 2, 45
Voigt model, *see* Rule of mixture

W
Wet layup, 34, 35, 55
Wet out issue, 6
Wrapping, *see* Mandrel lamination